Anolis Lizards of the Caribbean

Oxford Series in Ecology and Evolution
Edited by Robert M. May and Paul H. Harvey

Anolis Lizards of the Caribbean:
Ecology, Evolution, and Plate Tectonics

JONATHAN ROUGHGARDEN

Stanford University
Stanford, California

New York   Oxford
OXFORD UNIVERSITY PRESS
1995

Oxford University Press

Oxford   New York
Athens   Auckland   Bangkok   Bombay
Calcutta   Cape Town   Dar es Salaam   Delhi
Florence   Hong Kong   Istanbul   Karachi
Kuala Lumpur   Madras   Madrid   Melbourne
Mexico City   Nairobi   Paris   Singapore
Taipei   Tokyo   Toronto

and associated companies in
Berlin   Ibadan

Published by Oxford University Press, Inc.,
200 Madison Avenue, New York, New York 10016

Oxford is a registered trademark of Oxford University Press, Inc.

Library of Congress Cataloging-in-Publication Data
Roughgarden, Jonathan.
Anolis lizards of the Caribbean:
ecology, evolution, and plate tectonics /
Jonathan Roughgarden.
p.   cm. — (Oxford series in ecology and evolution)
Includes bibliographical references and index.
ISBN 0-19-506731-2
ISBN 0-19-509605-3 (pbk.)
1. Anoles—West Indies.
I. Title.   II. Series.
QL666.L25R68   1995   597.95—dc20   94-26777

1 3 5 7 9 8 6 4 2

Printed in the United States of America
on acid-free paper

Preface

Anolis lizards are the colorful green, gray, or brown lizards readily seen around houses, gardens, and woods in the West Indies. Anoles are active mainly during the day, as contrasted with the geckos also seen in and near houses, which are active primarily at night. Male anoles have a yellow, red, or white throat fan used in territorial displays and courtship; they extend and retract this fan in a special sequence to communicate with one another. Both males and females have toe pads, a sign of their arboreal habit. Together with students and colleagues, I have studied these animals on the islands of the eastern Caribbean during the last 20 years. This book offers a summary of why these animals have held our interest for so long and what we have learned.

Anoles typically perch on trees or bushes scanning the ground for insects. When a desirable insect appears, an anole runs and captures it, and hence an anole is called a "sit-and-wait" predator. Anoles occupy a niche filled by birds in North America and Europe. The West Indies lack most ground-feeding insectivorous birds such as robins, blue jays, and so forth; these are replaced by anoles. Birds are warm-blooded, which means they use metabolic energy to maintain the high body temperature required for effective locomotion. Lizards, in contrast, are cold-blooded, which does *not* mean they are cold but does mean they attain a high body temperature in a more elegant way—by basking directly in sunlight. Hence, an anole does not waste food to stay warm but uses all its food directly for maintenance, growth, and reproductive activities. Lizards are thus more efficient than birds, and a given food supply supports as many as 100 more lizards than birds. As a result, the abundance of lizards in the West Indies is phenomenal—one lizard per m^2 is typical. The total population size on a typical 400 km^2 island in the eastern Caribbean is therefore on the order of 10^8. This abundance implies that anoles are big players in the ecological communities found on Caribbean islands, and much of the ecosystem's energy and carbon flow through their populations.

Anoles are also big players in the zoological kingdom. The genus, *Anolis*, contains about 300 species, which makes it one of the largest genera of vertebrates.[1] These 300 species are distributed throughout Central America, northern South America, and the Caribbean islands, including the Greater Antilles (Puerto Rico, Hispaniola, Jamaica, and Cuba) and the Lesser Antilles (the small islands at the eastern margin of the Caribbean). Anoles are also found throughout the Bahamas. About half of the anole species occurs on Caribbean islands, the other half in Central and South America. These 300 species of anoles are about 5-10% of the entire present-day lizard fauna of the world.

The Greater and Lesser Antilles contain species unique to each island. For

---

[1] The official Latin name for a species consists of two words, the genus, which is capitalized, followed by the species itself, which is not capitalized, and both are italicized. The word "anole" is an informal common name for these animals.

example, the 11 *Anolis* species of Puerto Rico and its nearby cays are found only there. Hispaniola, adjacent to Puerto Rico on the west, has more than 35 *Anolis* species of its own, and none in common with Puerto Rico or Cuba. Closest to Puerto Rico on the southeast are the islands of Anguilla, St. Martin, and St. Barths. These are considered as a single "bank" because they are separated by shallow water. These three islands were united during the last glaciation period 15,000 years ago when the sea level was much lower. This bank has two species of anoles. These two species are found only there and nowhere else; they are not shared with Puerto Rico to its north or with the small islands still farther south. In this sense, then, the fauna on each of the Greater and Lesser Antillean islands is unique to the island.

A slight exception is that Jamaica has six native species, plus one invader from Cuba. Cuba itself, however, has a large native fauna of over 35 species that are not otherwise shared with Jamaica. In contrast, the Bahamas differ from the Greater and Lesser Antilles in *not* having any native species. The entire Bahamas appear to be under water occasionally and are recolonized anew from the adjacent Greater Antilles whenever they reappear.

From a scientific standpoint, the Caribbean islands with their *Anolis* lizards comprise a *system*. The islands offer miniature laboratories to study ecology and evolution. Indeed, this system is especially suited for the interdisciplinary subject of evolutionary ecology — the study of how an ecological context supplies the natural selection that drives evolution and of how evolutionary change among species in turn affects their ecological situation. Furthermore, it also appears that anoles may be used as "living strata" to aid in reconstructing the plate-tectonic origin of the Caribbean region, and perhaps also Central America. Together with other reptiles, lizards have a history that extends back over 100 million years, which is the period during which much of the Caribbean formed and major tectonic motion has occurred. These, then, are the scientific motivations for working with anoles on Caribbean islands.

Both our field studies and theoretical models have been directed mostly to the Lesser Antilles, simply because they are smaller and easier to understand than the Greater Antilles. Hence, this book is mostly about the small islands dotting the eastern margin of the Caribbean and, even more specifically, often aimed at the northeastern corner of the Caribbean where we have had the most experience.

Much of the research discussed in this book has been published previously in peer-reviewed scientific books or journals, sparing the need to reprint gory detail. Instead, the aim here is to synthesize, to compare what we thought we'd find with what we really did find, and to indicate where our understanding is still fragile. The theoretical aspects of Chapter 1, however, are new and not published elsewhere.

The computer programs developed in this book are written in a relatively new dialect, Scheme, of the venerable computer language, Lisp. Programs about an animal's ability to learn, to remember and to innovate are easy to express in this dialect. Also, web and tree-like data structures are found throughout biology, and all dialects of Lisp easily represent these data structures. Programs for all the calculations in the book are supplied on a diskette that can be purchased for a nominal price from the publishers. The programs are written in Scheme, with a few extras tossed in written in C and Pascal. Hopefully the interpretative nature of Scheme, together with the availability of the programs, will encourage a hands-on attitude to the theory presented in the book. Not only can the numerical value of a constant be changed on the fly but so can the actual structure of a model.

I am especially grateful to David Heckel, Steve Pacala, and John Rummel for their efforts during the initial period of this work, and to Steve Adolph, Roman Dial, Andy Dobson, Eduardo Fuentes, Lloyd Goldwasser, John McLaughlin, Ken Naganuma, Warren Porter, and Sharoni Shafir for important contributions at other stages. I acknowledge the efforts of students and others who have helped in the lizard censuses and related field activities. I also thank Mme. Flemming of St. Martin for allowing work to be carried out on her land for over 12 years, Vikki and Carl Leisegang for their excellent seamanship during the Anguillita experiments, and Mr. Bryan of West End, Anguilla for allowing the Anguillita experiments on his property. I am grateful to Ernest Williams for introducing me to *Anolis* many years ago, to Tom Schoener for valuable discussions, and to Bob and Judith May for hospitality during sabbatical visits to Princeton and Oxford. I am indebted to Jonathan Losos for an exquisitely detailed and thoughtful review of the entire manuscript. Finally, I thank my colleagues at Stanford for their interest and support over the years.

The field research reported here has been funded by the National Science Foundation, and the theoretical research primarily by the Department of Energy. Recently, research on the relation between adaptive computation and animal behavior has been supported by the Interval Research Corporation of Palo Alto.

This book is dedicated to the peoples of the Caribbean. It highlights biological treasures that are their inheritance.

Contents

Anolis Lizards of the Caribbean

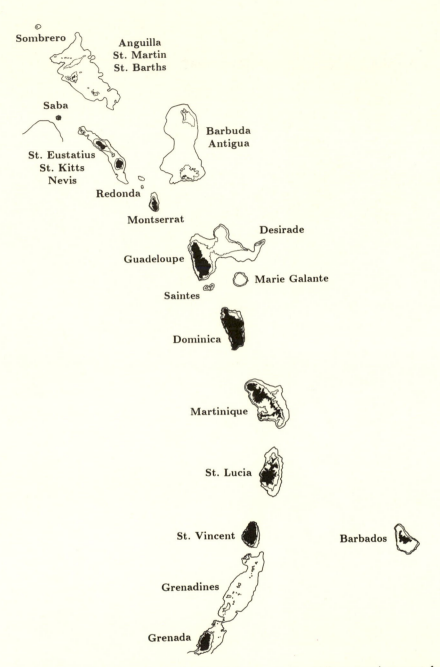

The Lesser Antilles. Outer contour is -50 m below sea level, middle contour is present-day sea level, and inner contour is 200 m above sea level; all land above 200 m is filled in black. The outer contour demarks banks that were united during times of the Pleistocene glaciation.

Plate I upper *Anolis oculatus*, males, from Dominica, illustrating various subspecies from around the island. The solitary anoles from Guadeloupe and Martinique also show this degree of subspecific differentiation within the island. **lower** *Anolis ferreus* from Marie Galante, shown mating to illustrate the largest sexual dimorphism in the Lesser Antilles. The larger individual is the male. This species is the exception to the biogeographic size rule for solitary species, primarily because of the large size of the males.

Plate II left (upper) *Anolis gingivinus*, male, from St. Martin. Although the larger species on a two-species island, its size is that of a solitary species, making it the exception to the biogeographic size rule for two-species islands. **(middle)** *Anolis pogus*, male and female, from St. Martin, the smaller species now found only in the hills in the center of the island, having become extinct on Anguilla. **(lower)** *Anoles schwartzi*, male, from St. Kitts, a small species from a two-species island in the bimaculatus group.

Plate II below (left) *Anolis aeneus*, male, from Grenada, a small species from a two-species island in the roquet group. **(middle)** *Anolis richardi*, male, from Grenada, a large species from a two-species island in the roquet group. **(right)** *Anolis bimaculatus*, male, from St. Kitts, a large species from a two-species island in the bimaculatus group.

1

The sentient forager

Suppose you awake tomorrow morning to discover you've become an anole. Will you know what to do? Will you live or die? Well, if you've watched enough anoles, you could imitate what you've seen. But it would clearly be better to know why an anole does what it does. Then you could adjust to new circumstances, including those you hadn't encountered before. This chapter is about how to prosper as an anole. It's not about all there is to an anole's life, but an important part—what to eat, how what you eat affects how fast you grow, and when to stop growing and to start reproducing.

This chapter pursues an ambitious program of biological synthesis. It predicts both the behavior and life history of an animal from what might be thought of as first principles. By combining physiological energetics, the ecological theory of optimal foraging, and the evolutionary theory of density-dependent natural selection, we predict the foraging behavior, growth, and life history of an *Anolis* lizard with qualitative realism and encouraging accuracy.

This chapter also addresses what many find to be a basic limitation of optimal foraging theory—its very plausibility for animals whose behavior is learned. Optimal foraging theory predicts how an animal *ought* to behave. One is then invited to suppose that natural selection brings about this optimal behavior, just as natural selection may shape a bird's wing in an optimal design. But behavior is rarely the expression of anatomy, or of a gene; it is the result of learning, and it changes with context. Therefore, we should inquire if there exists some behavioral rule that, when followed repetitively, leads an animal to behave optimally. If so, we may hypothesize that natural selection shapes an animal's instincts to follow this general rule so that behavior emerges in each circumstance that is optimal for that circumstance. What the animal has learned could then be identified with what is remembered while the rule is being followed. This chapter presents a simple rule that leads to optimal foraging behavior. A lizard can learn to forage optimally very quickly according to this rule—six minutes in simple cases. Indeed, this quickness of mind has induced me to offer lectures with the title "Are Lizards Smarter Than Deans?"

1.1 The optimal foraging distance

Because anoles usually perch at a spot in their territory scanning the ground for insects, it seems sensible to ask what determines the distance that an anole will run to capture a prey. Perhaps if a prey is detected beyond a particular distance, it will be ignored, while if within that distance, a chase will be triggered. This distance, to be called the "foraging cutoff distance," is important because it underlies how close in space lizards can be to one another while not eating each other's food. Also, the foraging cutoff distance determines the size of a lizard's foraging territory, which in turn largely determines how many lizards can be found in an area of habitat.

The approach to be used is "optimal foraging theory," introduced in the 1960's [105, 201] and developed extensively since then [60, 174, 204, 260, 261, 308, 310]. Optimal foraging theory begins with the choice of an optimization criterion, that is, some hypothesis for what it is that a lizard's foraging behavior optimizes.

One possible criterion is to minimize the average time used to capture an item. This criterion is reasonable in circumstances of high predation. A lizard is especially vulnerable when chasing a prey item; this is when a bird or snake can strike. So, the problem is to predict what a lizard must actually do to minimize the average time per prey item. Later in the chapter we will focus on another possible criterion, maximizing energy per time. Even if one prefers another criterion, minimizing time per item is a good place to begin because the calculations are always needed later on, and this criterion is sufficiently realistic by itself in some situations.

For our hypothetical lizard, time is used while waiting for prey to appear and while chasing down the item and then returning to the perch. Now here are the tradeoffs. If a lizard chases very distant items, its waiting time between the appearance of items is short, but its time spent pursuing the prey and returning to the perch is great. Alternatively, if a lizard pursues only nearby items, its waiting time for an insect to appear very nearby is long even though the time taken to catch it is short. Our problem then is to calculate a cutoff point that is the best compromise between the need for a short average waiting time and the need for a short average pursuit time.

To begin, suppose that a lizard can see what happens within 180° in front of it. If prey appear at random at any point, then the density of prey that appear between distance r and $r + dr$ from the lizard is

$$\rho(r)dr = a\pi r dr \tag{1.1}$$

where a is the abundance of prey in units of prey per m^2 per time. This is simply a uniform distribution in space transformed to polar coordinates. (If the lizard could see a full 360°, there would be an extra factor of 2 on the right-hand side of the equation.)

The waiting time is the inverse of prey abundance: Abundance is in units of prey per time and waiting time in units of time per prey. Now let r_c denote the cutoff foraging radius. This is the distance such that prey are chased if they are closer than this value and ignored if farther than this value. The total rate at which prey are appearing within r_c is the integral from 0 to r_c of the prey density

$$\int_0^{r_c} a\pi r dr \tag{1.2}$$

Therefore, the average waiting time between appearances of prey within this radius is

$$\frac{1}{\int_0^{r_c} a\pi r\, dr} \tag{1.3}$$

The next ingredient is the pursuit time. The lizard's sprint speed, v, is measurable. For a given prey distance, r, the time to chase it and return to the perch is $2r/v$. To calculate the average pursuit time, we need to multiply the probability that a prey is at distance r, times the pursuit time for the prey at r, integrated from 0 to r_c,

$$\int_0^{r_c} \frac{a\pi r}{\int_0^{r_c} a\pi r\, dr} \frac{2r}{v}\, dr \tag{1.4}$$

Now by adding the average waiting time to the average pursuit time we obtain the average time spent per item captured, T/I,

$$T/I = \frac{1}{\int_0^{r_c} a\pi r\, dr} + \int_0^{r_c} \frac{a\pi r}{\int_0^{r_c} a\pi r\, dr} \frac{2r}{v}\, dr \tag{1.5}$$

The expression for the average time per item, T/I, can be regarded as a function of r_c. The optimal foraging distance is therefore the value of r_c that minimizes T/I.

To obtain the optimal foraging distance, the equation for T/I can be simplified by doing the integrals explicitly, yielding

$$T/I = \frac{2}{a\pi r_c^2} + \frac{4r_c}{3v} \tag{1.6}$$

T/I clearly has a minimum because the first term becomes large when r_c approaches 0, and the second term becomes large when r_c approaches infinity. The minimum to T/I is then found to occur when the cutoff foraging distance is

$$r_c = \sqrt[3]{\frac{3}{\pi}\frac{v}{a}} \tag{1.7}$$

Notice that the foraging distance expands as the sprint speed increases, and shrinks as the food abundance increases, as one would intuitively expect. Perhaps not intuitive is the cube root; this indicates an optimal cutoff distance that is not especially sensitive to the precise numerical values of sprint speed and food abundance. A possible value of the sprint velocity for a lizard that measures 60 mm, not including its tail, is 1.5 m per second. A possible value of insect abundance is 0.003 prey per m^2 per s. The optimal foraging cutoff distance then works out to be about 8 m.

Now what could this result possibly mean for a lizard, even our hypothetical lizard? Will natural selection in a high-predation environment cause the evolution of a lizard brain that has a hard-wired trigger set for 8 meters? Unlikely, because the insect abundance varies from place to place, and the sprint velocity depends on the lizard's body temperature and perhaps on the difficulty of the terrain. Therefore, the value of 8 meters would be fine for some circumstances, but not others. Well, then perhaps natural selection has hardwired a lizard's brain with Equation 1.7, complete with the ability to take cube roots. If so, all a lizard has to do is assay the insect abundance and know its own sprint velocity to calculate the optimal foraging cutoff distance for itself. Yet this clearly stretches credulity.

Moreover, perhaps the foraging model has hopelessly oversimplified the way a lizard forages. The lizard may not be under strong predation pressure, implying that minimizing T/I is a poor choice as the optimization criterion. The lizard may be more interested in the caloric or nutritional value of the prey than in the time it takes to run it down, and it may not be as good at chasing prey as assumed, for prey often do escape. And so forth.

Two issues thus have to be faced. First, how do we take *learning* into account? Can we conceive of a simple behavioral rule to describe how a lizard might think and examine whether the behavior that results from this rule eventually leads the lizard to do what we may know, *a priori*, is optimal for it? That is, can even our hypothetical lizard learn to forage optimally without the need to calculate cube roots in its head? Second, how can we add enough about the way lizards do forage to make the model more than a theoretical curiosity? The problem here is not to shove all conceivable detail into the model, resulting in a supermodel that covers all imaginable situations. Instead we will seek a model that is realistic enough for our purposes, and yet still understandable. Both these issues will be addressed in sequence. In the next section, we consider learning for our hypothetical lizard. Then the bulk of this chapter is devoted to tailoring an optimal foraging model for *Anolis* lizards. This special *Anolis* foraging model will thereafter be available for many uses in the remainder of the chapter.

1.2   Learning to forage optimally

Can a forager that is initially naive about what to forage upon somehow learn how to forage optimally? Can nonoptimal behavior slowly develop into optimal behavior?

1.2.1   Rule of thumb

Here's what we'll imagine goes on during a foraging episode taken from the life of a lizard:

- The lizard watches from its perch for prey. When one appears, it notes the number of seconds it waited for this appearance and the x and y coordinates of the item. It places a list whose components are the waiting time, x coordinate, and y coordinate in its memory as the most recent event that has happened to it.

- With this observation in mind, the lizard thinks about what to do. It estimates the distance to the prey using the x and y coordinates. It also conceives of two possible actions: one being how to catch the prey, the other how to ignore the prey. It places a list whose components are the distance to the prey, and the two possible courses of action, in its memory as the most recent event. At this time the lizard makes and stores in its memory two executable functions; it doesn't actually execute either of them, but it does make them up.

- With these two possible actions in mind, the lizard evaluates which is best. Based on the experience accumulated in its memory, the lizard anticipates what its time per item captured (T/I) will be if it pursues the prey, and what T/I will be if it ignores the prey. Then the lizard decides in favor of the action anticipated to bring about the lower T/I. It then places in memory a list with the recommended decision and the projected outcomes as its most recent event in memory.

• Now the sentient lizard acts. It executes the recommended function, and stores the outcome in its memory. Specifically, it stores its cumulative waiting time, which is the previously accumulated waiting time incremented by the time spent waiting for the particular item under consideration. Also, if it has chosen to pursue the prey, then the cumulative pursuit time is incremented by that used to run to the prey and back, and the cumulative number of items caught is incremented by 1. Every pursuit is assumed to succeed. Alternatively, if it has chosen to ignore the prey, then the cumulative pursuit time and number of items caught are left unchanged. Then it begins looking again for prey, and the loop repeats.

The lizard begins life without experience, but with a willingness to eat the first item that appears. Thereafter, repetitions of the loop above lead to the accumulation of experience in the form as represented by the remembered cumulative catch I, and cumulative elapsed time, T.

The loop described above can be thought of as a rule of thumb [173, 162]. The rule is, at each appearance of an insect, to do which ever action leads to a lower projected T/I. That is, when a new insect appears, the projected T/I if the item is pursued is

$$(T/I)_p = \frac{T + t_w + t_p}{I + 1} \tag{1.8}$$

and the projected T/I if the item is ignored is

$$(T/I)_i = \frac{T + t_w}{I} \tag{1.9}$$

where t_p is the anticipated time to pursue the insect and to return to the perch, and t_w is the time since the previous insect appeared. T is the total elapsed time through whatever was done with the previous insect, and I is the total number of insects caught up to now. The rule is to do whichever action leads to

$$\min[(T/I)_p, (T/I)_i] \tag{1.10}$$

After the action, the T and I are updated accordingly. You'll note that the loop described above is a bit more elaborate than this simple rule; this is because I'd like to imagine enhancing the loop to include discovery and other acts of cognition in future research.

On following this rule of thumb, a regularity in the foraging behavior eventually emerges. The experienced lizard consistently ignores prey beyond some distance and takes prey within that distance. That is certainly comforting, but the big question is: Does the foraging cutoff distance that the lizard develops actually *equal* the cutoff distance predicted by optimal foraging theory? Surprisingly (to me), it does. This hypothetical lizard does learn to be optimal.

1.2.2 A computer program

The next two sections offer a digression on how to implement a rule-of-thumb computer program in Scheme, a dialect of Lisp. Readers may wish to skip to Section 1.2.4 "Our lizard learns fast." In these next two sections the focus is solely on software technology, and no new biology is introduced.

Scheme has the unusual property of allowing functions to be created during program execution: A program can write itself as it runs. In traditional computer languages, such as Pascal, C, and Fortran, a fundamental distinction is drawn between the program's instructions (code) and the variables and constants that these instructions act on (data). In such languages it may be possible to create variables dynamically, i.e., while the program is running, even though such variables were not known to be needed before the program was started up. The new and malloc() function calls in Pascal and C serve this role. They gather some space in memory for the program's use even though this memory wasn't originally budgeted. Still, the existing set of program instructions is what operates on these newly created variables; the program's inherent capabilities remain static and never exceed what the program had to begin with. In contrast, instructions in a Scheme program can expand as the program runs, so that the program's capabilities actually expand. This feature seems ideal for exploring the role of learning in animal behavior.

Moreover, Scheme is usually implemented as an interpreter, somewhat like Basic. This means that programs and data can be changed immediately, to satisfy curiosity, to follow a hunch, or just for fun. In the program that follows, if you don't like something, just change it.

Now, as promised, the program to carry out simulated learning in our hypothetical lizard is detailed. It is called learn.scm on the diskette.

Hints on reading Scheme

Most readers will not be familiar with the syntax of a Lisp program. Still, just by looking over the program below you'll get an idea of what's going on, and here are some hints to help you do this (cf. [1, 102, 104, 337, 158]).

When Scheme is started up from a terminal, it listens to you and responds to anything you type. Whenever you type something, it trys to evaluate what you type. If you type the number 7, it replies with 7 because a number simply evaluates to itself. But if you type x, it replies that x is unknown to it.

When you give Scheme an expression to evaluate that begins and ends with parentheses, it assumes that what's in the parentheses is the name of a function followed by the arguments to that function. For example, if you type (+ 1 3), Scheme responds with 4. Here "+" is the name of the function that adds two numbers, and what are to be added are viewed as the arguments to this function. Another example is the function that acquaints Scheme with the names of variables. If we type (define x 7), then Scheme uses the define function with x and 7 as arguments. define causes computer memory to be allocated to the program and the value 7 to be stored at a computer memory location named x. Thereafter, if you type x, Scheme replies with 7 because now Scheme knows about x.

The way a brand new function is manufactured is to package one or more statements together and to assign a name to the package. The statement packaging is done with a function called lambda, and the package made by lambda is then assigned a name with define. Thus, to manufacture a function that doubles a number, to be called double, one types:

(define double(lambda(x)(+ x x))).
To see what this means, start in the inside and work out. The (+ x x) is where
x is doubled; lambda packages this expression into an entity that can be given a
name and stipulates that x is understood as an argument. define is what actually
assigns the name. Once double's definition has been given to Scheme, it can be
used; for example, typing (double 2) will now cause Scheme to reply with 4.

Of course, a program does not have to be typed anew each time. It can be
written outside of Scheme with a text editor and then loaded in at the start of the
Scheme session using the load function. And once loaded, the program can be
changed at will. The output of the program can be recorded permanently (on a
disk) with the transcript-on function. The transcript can then be printed as is
or incorporated into other documents with a text editor.

Concerning syntax, an expression does not have to fit on one line. Indeed, most
expressions, from the beginning to ending parenthesis, are spread across several
lines. Also, anything on a line after a semicolon is ignored. Hence, comments can
be embedded in the program by placing them on lines that begin with a semicolon.
Scheme does not distinguish case, so x and X refer to the same variable.

As you've noticed, arithmetic operations use prefix notation. For example,
(+ 1 1) evaluates to 2. Similarly, (* 2 2) evaluates to 4, (/ 21 7) to 3, and
exponentiation is carried out with expt so that (expt 2 8) yields 256. Numbers
in Scheme include the familiar integers and the real numbers (floating-point numbers). Scheme can also represent rational numbers (exact fractions). The exact
fraction, 1/3, is meaningful in Scheme. While the exact representation of rationals
is a standard feature of Scheme, some implementations fail to include it.

When you look over the program you'll notice that lines are often indented.
The indentations illustrate block structure in the program. The let function is
used to construct a block. Local variables are defined within a let function and
have scope extending only to the statements contained within the let construction.
Also, Scheme has a powerful looping syntax implemented with the do function.
The contents of a do loop also form a local block.

The compound data structure typically used in Scheme is the list. It is simply a
sequence of data bounded by parentheses, such as (1 2 3). This list can be given
a name with define, just as other entities can. But to distinguish a list from a
function call, it must be preceded by a single quote mark '. Thus, to name a list one
would type (define data '(1 2 3)), whereupon data now has as its value the
list consisting of the numbers, 1, 2, and 3. You could now type data to Scheme, and
it would reply with (1 2 3) as its answer. If the ' were not used, then the attempt
to define data would produce an error, because Scheme would try to evaluate the
1 as though it were the name of a function. A list can also be manufactured with
a function called list that packages its arguments together into a list. Indeed, an
equivalent way to define data is (define data (list 1 2 3)). The functions
to extract individual elements from a list are called car and cdr. car pulls out
the first element of a list, and cdr yields the remainder of a list, excluding the first
element. So, these are used in combination. To get the second element of a list, use
a cdr to discard the first element, then a car. For example, (car (cdr data))
yields 2. Because this idiom is used so often, the compound operation cadr was
devised. (cadr data) also yields the second element of the list data, which is
2 in this example, and is synonymous with (car (cdr data)). Similarly, the
compound operator caddr returns the third element of a list.

Finally, you'll notice the symbol #t, which stands for True. False is represented
as #f and sometimes also as (). The symbol () is also used for a list without any

Table 1.1 Parameter definitions.

```
(define abund (/ 120 (* 12 60 60)))
     ; 120 prey per sq m per 12 hr, converted to seconds
(define svl 60)
     ; snout-vent-length in millimeters, a typical size
(define mass (* 3.25e-5 (expt svl 2.98)) )
     ; fresh weight in grams, given the svl
(define velocity (* 0.75 (expt mass 0.35)) )
     ; maximum sprint in meters per second, given the mass
(define xmax 12)
     ; meters, study area is 2 times xmax by 1 times ymax
(define ymax 12)
     ; with lizard located at the origin
(define quan (* abund 2 xmax ymax))
     ; prey in study area, must be less than 1.0
(define nbig 1000)
     ; reciprocal of increment for random numbers in [0 1)
```

---

elements, the so-called "empty list." By convention, Boolean variables are named with a question mark as the last character in the name. For example, attack? would be a typical name for a variable whose value is either #t or #f. Also, by convention, a function that changes the value of some variable that has already been defined is given a name with an exclamation mark as the last character. For example, set! reassigns a variable to a new value, as in (set! x 8). Thus, even though x was originally defined as equal to 7, its value has now been changed to 8. If you type x, Scheme now replies with 8.

Biological parameters

We begin by entering some biological parameters that characterize the abundance of insects, length of the lizard, mass of the lizard, and sprint velocity of the lizard, as shown in Table 1.1.

We also define the size of the area being scanned. Further definitions include quan, the probability that an insect appears each second somewhere throughout the entire area being scanned; it is computed from the insect abundance times the area being scanned. The last definition, nbig, is used when generating random numbers.

The predicted foraging radius

Using these biological parameters, the optimal foraging cutoff distance, from Equation 1.7 derived in the last section, is entered as:

```
(define r-optimal
    (expt (/ (* 3 velocity) (* 3.14159 abund)) 1/3 ) )
```

At this point, just type r-optimal to the interpreter and it responds with 7.91, which is the optimal foraging cutoff distance in meters for these parameter values. The question to be solved with the simulation is whether this radius predicted by optimal foraging theory ever equals the cutoff radius developed by an experienced lizard.

What the lizard remembers

The lizard's memory will be a list of foraging episodes accumulated through life. Each episode is itself a list consisting of an observation, proposition, decision, and result. Each new episode is prepended to this list, so the head of this list is the most recent episode. The memory is initialized to an empty list, to be called birth,

```
(define birth ()); start life with no memory
```

What the lizard sees

Now we type in the look function, shown in Table 1.2, that models the lizard when it is sitting at its perch looking for prey. look returns a prey observation, which is the list

```
(waiting-time x-coordinate y-coordinate)
```

where x-coordinate is in [−xmax xmax] and y-coordinate is in [0 ymax], and waiting-time is in seconds.

look uses Scheme's built-in random function to deliver the random numbers. A call to (random n) returns an integer between 0 and n−1. look works by looping each second until a prey item appears, and then assigns random coordinates to it. Calling (/ (random nbig) nbig) returns a random number between 0 and 1. The probability that an insect appears each second is quan. So, if the random number is less than quan then an insect has appeared. The do loop then evaluates to a list consisting of the number of iterations (seconds) it has taken for the insect to appear, together with an x coordinate chosen at random between −xmax and xmax and a y coordinate chosen at random between 0 and ymax.

What the lizard thinks

Now we come to the function that separates our lizard from some lower organism— the ability to think.

As Table 1.3 shows, think takes as its argument an observation: This is the object that was produced previously by look. think returns a proposition, which is the list:

Table 1.2 The look function.

```
(define look (lambda ()
    (do ( (seconds 1 (+ 1 seconds)) )
        ( (<= (/ (random nbig) nbig) quan)
          (list seconds
                (- xmax (random (+ 1 xmax xmax)))
                (random (+ 1 ymax))) )
        () ) ))
```

Table 1.3 The think function.

```
(define think (lambda (observation)
  (let* ( (waiting-time (car observation))
          (x-coordinate (cadr observation))
          (y-coordinate (caddr observation))
          (distance-to-prey
                (sqrt (+ (* x-coordinate x-coordinate)
                         (* y-coordinate y-coordinate)))) )
        (list distance-to-prey
              (lambda (result)
                 (let* ( (time-pursuit-total (car result))
                         (time-waiting-total (cadr result))
                         (items-total (caddr result))
                         (observations-total (cadddr result)))
                    (list (+ (/ (* 2 distance-to-prey)
                            velocity)
                            time-pursuit-total)
                          (+ waiting-time time-waiting-total)
                          (+ 1 items-total)
                          (+ 1 observations-total)) ))
              (lambda (result)
                 (let* ( (time-pursuit-total (car result))
                         (time-waiting-total (cadr result))
                         (items-total (caddr result))
                         (observations-total (cadddr result)))
                    (list (+ 0 time-pursuit-total)
                          (+ waiting-time time-waiting-total)
                          (+ 0 items-total)
                          (+ 1 observations-total)) )) )) ))
```

```
(distance-to-prey proposal-to-pursue
 proposal-to-ignore)
```

where (proposal-to-pursue last-result) and (proposal-to-wait last-result) offer alternative proposals for updating the cumulative statistics on foraging activity according to whether the prey has been pursued and caught, or ignored. These proposals are themselves functions; each is called with a previous result as its argument, and a new result is returned. A result is defined as a list of four cumulative statistics:

```
(total-pursuit-time total-waiting-time
 total-items-caught total-observations)
```

The first collection of statements in think unpacks the x and y coordinates from the observation; the distance to the object is then calculated. Next, the list that this function returns is assembled: See the call to list followed immediately by distance-to-prey. The last two entries in the list returned by this function are themselves functions; note that each is a collection of statements packaged together by lambda. These functions are not given explicit names because we know where they are and don't need to refer to them by name. Even though presently unnamed, they are perfectly executable. These functions, although representing possible courses of action for the lizard, actually do something mundane: They merely update some statistics.

How the lizard decides

Armed now with two proposed courses of action, the lizard must decide which is best. As shown in Table 1.4, decide is called with a proposition and with the presently accumulated experience as arguments. The lizard's experience is the list of all the events that have happened to it and is named memory in the function shown in Table 1.4. decide returns a list consisting of three elements: a boolean value of true or false to indicate a decision in favor or against pursuing the insect, a projected T/I if it were to pursue, and a projected T/I if it were to ignore the prey. Thus, decide returns a decision, thumbs up or down, together with the reasons behind the decision. The first set of statements in decide simply unpacks the relevant information from the arguments. In particular, the function that catches the insect is extracted from the proposition, and locally named pursue, and the function to ignore the insect is locally named ignore. Then both these functions are executed, and the T/I that results from each is determined, where T is the sum of the pursuit time and waiting time, and I is the sum of the items caught. The decision and the T/I resulting from the proposed actions are assembled into a list for return. The list (#t 0 0) is returned when decide is called with an empty list, (), as the initial memory.

The time for action

After all this thought and decision-making, act in Table 1.5 is anticlimactic: The lizard simply follows its own advice and carries out the action it has recommended to itself.

Specifically, act takes takes a decision, proposition, and memory as arguments and returns a result. As in the other functions, the first few statements

Table 1.4 The decide function.

```
(define decide (lambda (proposition memory)
   (if (null? memory) '(#t 0 0)
       (let* ( (episode (car memory))
               (result (cadddr episode))
                (pursue (cadr proposition))
                (ignore (caddr proposition))
                (evaluate (lambda (result)
                   (let* ( (time-pursuit (car result))
                           (time-waiting (cadr result))
                           (items (caddr result)) )
                      (/ (+ time-pursuit time-waiting) items) )))
                (t/i-if-pursue (evaluate (pursue result)))
                (t/i-if-ignore (evaluate (ignore result))) )
             (list (< t/i-if-pursue t/i-if-ignore)
                   t/i-if-pursue
                   t/i-if-ignore )))))
```

extract the information from the argument lists; also the initial condition is considered as a special case. Then, on the recommendation as stored in the variable conclusion?, act executes either pursue or ignore.

Combining life's parts

The final function, called live, is shown in Table 1.6.

It orchestrates the lizard's life as a sequence of foraging events. live is called with two arguments: an integer to indicate how many episodes to live, and the name of the master list of everything that has happened to the lizard. Usually, this is the empty list defined as birth. It could also be a list that already has

Table 1.5 The act function.

```
(define act (lambda (decision proposition memory)
   (let* ( (result (if (null? memory) '(0 0 0 0)
                       (let ( (episode (car memory)) )
                          (cadddr episode))))
           (conclusion? (car decision))
           (pursue (cadr proposition))
           (ignore (caddr proposition)) )
        (if conclusion? (pursue result)
                        (ignore result)) ) ))
```

Table 1.6 The `live` function.

```
(define live (lambda (lifespan initial-memory)
    (do ( (current-memory initial-memory)
          (sightings 0 (+ 1 sightings)) )
        ( (= sightings lifespan) current-memory)
        (let* ( (observation (look))
                (proposition (think observation))
                (decision (decide proposition
                                  current-memory))
                (result (act decision
                      proposition current-memory))
                (episode (list observation
                      proposition decision result)) )
          (set! current-memory
                  (cons episode current-memory)))) ))
```

something in it; for example, successive calls to `live` using the output of the first call as the input to the second call would extend the lizard's life. `live` returns the list consisting of everything that has happened to the lizard during the period when `live` was active, all concatenated to the initial list. `live` simply calls `look`, `think`, `decide`, and `act` in that order, while passing the output of each as the argument to the next. After all four functions have been executed, their output is bundled into an `episode` and the episode prepended to the lizard's cumulative memory.

1.2.3 Trying it out

The program is now complete. It is executed for a life span of two episodes by typing, for example,

```
(define shortmemory (live 2 birth))
```

This invocation of `live` means that the lizard experiences two foraging episodes starting with the empty list, `birth`, as its initial memory. `live` returns the cumulative memory list after the two foraging episodes have transpired, and this new memory is here given the name `shortmemory`.

So, at this point the Scheme interpreter has in its possession a list consisting of two foraging episodes in our hypothetical lizard's life. How, you then ask, do we get Scheme to tell us what happened? We just type `shortmemory` and Scheme displays `shortmemory`; the printout may not look great, but *all* the information is there. Specifically, what Scheme types out to us is shown in Table 1.7. I've taken the liberty of indenting and editing the printout into the form above for clarity.[1]

---

[1]Table 1.7 is the output from `learn.scm` using version 6.1 of MIT-scheme. More recent versions and other implementations do not refer to the procedure by its hexadecimal address, but by some other label such as procedure-1, procedure-2, and so forth.

Table 1.7 Contents of shortmemory.

```
((
    (1 2 7)
    (7.28 #[PROCEDURE #x6CE43] #[PROCEDURE #x6CE41])
    (#T 11.60 13.10)
    (21.20 2 2 2)
  )
  (
    (1 8 0)
    (8.00 #[PROCEDURE #x6CC22] #[PROCEDURE #x6CC20])
    (#T 0 0)
    (11.10 1 1 1)
))
```

The overall memory list consists of two episodes, with the later on top and the earlier at the bottom. Within the first episode, look spotted an insect after 1 second at (x, y) coordinates of (8,0). think established that this insect was 8 meters away, and produced two functions, one known to the computer at machine address #x6cc22 and the other at machine address #x6cc20 (Scheme's notation for hexadecimal). These are the executable functions that correspond to pursuing or ignoring the particular item being considered in this episode. Then decide concluded that the first of these functions, the one to pursue the prey, should be carried out. It decided this simply because the lizard is assumed to pursue the first prey it ever sees. Then act does chase the insect and reports that the total pursuit time is now 11.10 seconds, the total waiting time is 1 second, the total number of items caught is 1, and a total of 1 prey have been seen.

The second episode is much the same. After 1 second a prey was spotted at (x, y) coordinates of (2, 7). This insect was 7.28 meters away, and functions to execute if it is to be pursued, and if it is to be ignored, were proposed. decide recommended pursuing the insect because it projected that T/I will be 11.60 seconds per item if it is chased and caught and 13.10 seconds per item if ignored. Therefore act did chase the item and reported that the total pursuit time is now 21.20 seconds, the total waiting time is 2 seconds, the total number of items caught is 2, and a total of 2 prey have been seen.

1.2.4 Our lizard learns fast

Because a complete report of everything that has happened to the lizard during its life, especially for a long life, is not of much interest, the learn.scm program on the diskette includes another function, recall, that traverses the complete memory list and produces a summary. To use this with a long life, say of 200 episodes, type

```
(define longmemory (live 200 birth))
```

Then type

```
(recall longmemory)
```

and a summary of the 200 episodes will appear. As already mentioned, each episode is, by definition, a prey sighting. Some sightings result in pursuits, while in others the prey is ignored.

Figure 1.1 offers an illustration of 200 episodes obtained in this way. It illustrates that our hypothetical lizard does learn how to forage optimally. At first the lizard takes prey beyond the optimal cutoff distance. But by the 60th episode it has learned the optimal cutoff distance to the resolution of this simulation. Notice the lizard seems to learn exactly where the cutoff is, because after the 60th episode it ignores prey even slightly beyond the optimal cutoff, and pursues prey even slightly within the optimal cutoff. Further runs with the program, not illustrated, show that the time required to learn where the optimum cutoff is depends on the distance of the first prey item caught. If this is far from the cutoff, more experience is needed to converge to the optimal decision. Still, 100 episodes have been enough in the runs tried so far. The 100 episodes of experience, corresponding to 100 prey sightings, typically involve about 65 prey that are ignored and 35 that are pursued. The average time per item, T/I, stabilizes at about 10.8 seconds per item. The 100 episodes take place during a total elapsed time of about 375 seconds, i.e., just over 6 minutes. Thus our hypothetical lizard learns very quickly.

What's happening mathematically when the lizard has "learned" the optimal cutoff distance is that its realized T/I has approached the theoretical T/I of Equation 1.6 evaluated at the r_c given by Equation 1.7. The behavior of this learned lizard follows the marginal value theorem of optimal foraging theory [60].

Other metaphors for learning to forage

Genetic programming The rule of thumb exhibited here is one metaphor that shows how optimal behavior can be attained developmentally. Another is supplied by genetic programming [147, 168, 169]. The idea here is to "evolve" computer programs. An initial "population" of computer programs is somehow produced. The effectiveness of each program in this population is tested with respect to some application, and then the most effective programs are allowed to "breed." Breeding means that the good programs actually exchange code fragments leading to new computer programs, and the programs that are not allowed to breed are discarded. A new generation of computer programs is thereby synthesized, and the effectiveness of each program in this new population is then tested, and so forth, generation after generation. Koza [168, 169] has shown that computer programs that solve classic problems in many disciplines can be manufactured in this way.

Recently, the technique of genetic programming has been applied to the problem of a lizard's optimal foraging [170]. The problem is to see if genetic programming can lead to a rule that makes the lizard forage optimally, i.e., to chase prey up to but not beyond the optimal cutoff distance. Notice how this metaphor is quite different from the rule-of-thumb metaphor. The genetic programming metaphor is to develop a foraging rule from elementary parts using a selection process, while in the rule-of-thumb metaphor the rule is not decomposed into more elementary subunits. In both metaphors though, the objective of the rule is to bring about optimal foraging.

In this test case of genetic programming applied to optimal foraging, the "elementary abilities" of the animal were assumed to be the real-valued operations of addition, subtraction, multiplication, division (with 0 returned if the denominator is 0), an exponentiation function with two arguments whereby the absolute value of the first argument is raised to the second argument, and a branch-if-less-than-

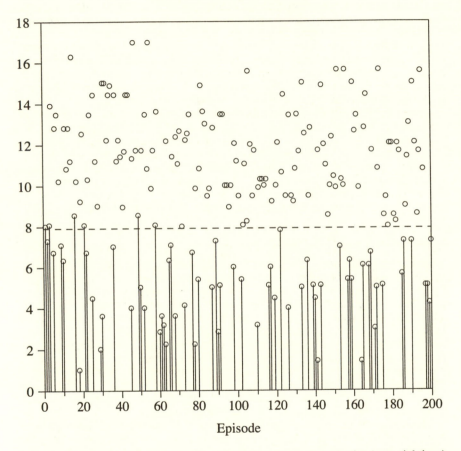

Distance of Prey from Lizard

Episode

Fig. 1.1 Distance in meters of prey from lizard in consecutive episodes (prey sightings). A vertical rule is drawn to any prey that is pursued. The optimal prey cutoff distance is shown as a horizontal dashed line. An "optimal forager" does not pursue prey farther than the dashed line. Illustration shows that the lizard's decisions approach optimality as time passes, indicating the effect of learning and experience. By the 60th episode, the lizard has learned where the optimal cutoff is and makes no further mistakes.

or-equal instruction. The setup is that an insect appears at position (x, y) in the 180° area visible in front of the lizard's perch at (0,0). The rule for the optimal foraging strategy in this context is $\text{Sig}(\sqrt[3]{\frac{3}{\pi}\frac{v}{a}} - \sqrt{x^2 + y^2})$ where Sig is the sign function. For any given insect, if this expression is +1 the insect should be chased and if -1 should be ignored. The question is whether genetic programming using the six elementary abilities just mentioned can produce a rule that is equivalent, or nearly so, to this optimal rule.

Starting with a randomly generated initial population of programs, some of which represented truly terrible foragers and others merely mediocre foragers, and using a population each generation of 1000 individual programs, the following "individual" was the best forager within the population at generation 60:

```
(+ (- (+ (- (expt a -0.9738) (* (expt x x) (* x
a))) (* (+ v a) (y)) (+ 0.7457 0.338898)))) (expt y
x)) (- (- (expt a -0.9738) (expt -0.443604 (- (- (+
(- (expt a -0.9738) (expt -0.443604 y)) (+ a x))
(expt y x)) (* (* (+ x 0.0101929) a) x)))) (* (expt
y y) (* x a))))
```

This foraging rule, though cobbled together in a seemingly haphazard fashion, does a very good job. In a simulated foraging session that happened to involve 17,256 insects, this "evolved" forager agreed with the optimal forager in all but 19 insects. Its foraging region was approximately a semicircle with radius r_c, even though it had some peculiar kinks. The population of programs at generation 60 included several other evolved rules that were nearly as good. So, we have another metaphor for how optimal foraging may be attained; as the result of a selection process on various combinations of elementary abilities.

The significance of this metaphor is twofold. (1) It suggests that optimal foraging strategies can be attained by cobbling together elementary abilities as a result of a selection process. (2) This, in turn, suggests that an interesting future step is to identify and to assemble one or more sets of biologically natural elementary abilities, and to see if a selection process on these also leads to an optimal forager.

Neural networks A third and comparatively well-developed metaphor for learning comes from neural networks (cf. [189, 143]). A neural net consists of nodes and interconnections and can be classified by criteria including the number of levels of nodes, whether the inputs are binary or continuous, and the properties of the nodes themselves. A net is typically "trained" by presenting it with inputs and adjusting the weights on each connection.

From a biological standpoint, a key distinction is whether the training is "supervised." Most of the literature pertains to the supervised situation in which the net is trained by comparison with a set of examples. However, "unsupervised" training is appropriate here, and specifically "reinforcement learning" algorithms[2] [362, 387, 386]. An *Anolis* lizard hatches from an untended egg about two months

---

[2]The sentient forager of this chapter compares only the immediate reward with experience. Interestingly, there are also two reinforcement-learning architectures (adaptive heuristic critic and Q-learning) that apply when future impacts of decisions and actions are also considered. These architectures are relevant to whether an animal can learn to farm a private garden in an optimal way, as perhaps occurs in some social insects.

after being deposited in the ground and begins life without parental care. There-fore, an anole's learning reflects only its own interaction with the environment, its own trials and errors. I emphasize, though, that an optimally foraging neural net has not yet been developed; instead, I am conjecturing that it can be.

The significance of this metaphor is also twofold. (1) While the neurobiological basis of a complex behavior like foraging is unknown, and while neural nets are not intended as accurate mimics of neurons and synaptic connections, the training of a neural net still offers a relatively nuts-and-bolts physiological metaphor for how optimal foraging can develop. (2) This metaphor should be extensible to learning more than the optimal cutoff radii, given the wide use of nets in pattern recognition applications. It should be possible to train the net to identify various kinds of prey according to their energy contents and probabilities of escape. Therefore, the lizard need not inherit the knowledge of a naturalist but can learn how to identify its prey during the course of its foraging.

In summary, three metaphors for how an animal can learn to forage optimally can be ordered in their degree of mechanistic detail. A rule of thumb just stands by itself, a rule obtained by genetic programming is composed of elementary behav-ioral abilities, and a rule brought about through training a neural net suggests an eventual neurobiological setting. All three metaphors show how optimal foraging can be learned, and learned quickly.

1.3 Energy as a criterion

Our hypothetical lizard has illustrated an idea, that a simple behavioral rule leads to behavior that coincides with what is predicted by optimal foraging theory. This hypothetical lizard should not be compared with any real lizard that I know of, and certainly not with a West Indian *Anolis* lizard. We're not yet modeling, even in an oversimplified way, what is of interest in *Anolis*. Specifically, we know that West Indian anoles receive little predation pressure relative to continental lizards [217], and the criterion of minimizing foraging time per prey item seems inappropriate. The value of 8 meters for the foraging radius of a lizard seems too large for an anole. Moreover, prey are not all the same. It's obvious that an anole will travel farther to catch a large juicy cricket than a small peppery ant. So, let's modify this model to address the situation of a West Indian *Anolis*. The model will continue to offer a simplification of behavior, perhaps overly so, but it will eventually be relevant enough to warrant testing.

The foraging strategy that would seem most appropriate for an anole is one that maximizes the rate at which energy is being accumulated. So, let's see what an anole should do to maximize this criterion. Actually, this task is only the first of three steps in introducing what seems an apt degree of realism to this foraging theory. The second step will be to consider energy-maximizing strategies with explicit provision for prey escape, and the final step will be to consider multiple prey.

1.3.1 Formulation with one prey

Energy maximization with one prey type is a straightforward extension of the theory for time minimization. Recall from Equation 1.6 that the average time per item T/I is

$$T/I = \frac{2}{a_1 \pi r_1^2} + \frac{4r_1}{3v} \tag{1.11}$$

where the first term is the waiting time per item and second term is the pursuit time per item. To anticipate our later notation when more than one prey type is involved, we let r_1 be the foraging cutoff distance for prey type 1 and a_1 the abundance of prey type 1. For now, we consider only one prey type—a "typical insect." The lizard's sprint velocity is still denoted as v. Next, we also need to know the average energy obtained per item, E/I. Then the *rate* of energy accumulation, which is a criterion we desire, will be found as E/I divided by T/I; the I divides out leaving E/T. Now the energy per item, E/I, is computed by separately considering energy expended while waiting and while pursuing the prey. Let e_w be the energy expended per time while sitting perched and looking for prey items, e_p be the energy expended per time while pursuing a prey item, and e_1 be the energy content of an insect of type 1. Then the average energy per item is

$$E/I = e_1 - \frac{2}{a_1 \pi r_1^2} e_w - \frac{4r_1}{3v} e_p \tag{1.12}$$

The first term is the energy yield on capturing the insect, the second term is average time spent waiting for the insect times the energy expended per unit waiting time, and the third term is the average time spent pursuing an item times the energy expended per unit pursuit time. The overall rate of energy accumulation is therefore

$$E/T = \frac{e_1 - \frac{2}{a_1 \pi r_1^2} e_w - \frac{4r_1}{3v} e_p}{\frac{2}{a_1 \pi r_1^2} + \frac{4r_1}{3v}} \tag{1.13}$$

Our task now is to find the foraging cutoff distance, r_1, that maximizes E/T. Mathematically, it's reassuring to note that if e_w equals e_p, then maximizing E/T is equivalent to minimizing T/I, as before. Thus, in a time-minimizing problem, waiting and pursuit time count equally—time is time no matter how it's spent—whereas in an energy-maximizing problem, waiting and pursuit time count differently because pursuit time is more expensive than waiting time.

E/T as a function of r_1 has one maximum. The y-intercept is $-e_w$, and the graph asymptotes at $-e_p$ as r_1 tends to ∞. On differentiating E/T with respect to r_1, setting the derivative equal to zero, and simplifying, we obtain

$$\pi Q r_1^3 + 6(e_p - e_w)r_1 - 3e_1 v = 0 \tag{1.14}$$

where

$$Q \equiv a_1 e_1 \tag{1.15}$$

The appropriate root to this cubic is

$$r_1 = \frac{-\frac{2(e_p - e_w)}{\pi Q} + \left(\frac{3e_1 v}{2\pi Q} + \sqrt{\left(\frac{3e_1 v}{2\pi Q}\right)^2 + \left(\frac{2(e_p - e_w)}{\pi Q}\right)^3} \right)^{2/3}}{\left(\frac{3e_1 v}{2\pi Q} + \sqrt{\left(\frac{3e_1 v}{2\pi Q}\right)^2 + \left(\frac{2(e_p - e_w)}{\pi Q}\right)^3} \right)^{1/3}} \tag{1.16}$$

Table 1.8 Insect Characteristics

```
(define insect-mass (lambda (insect-length)
    (* 0.03 (expt insect-length 2.5)) ))
    ; dry weight in mg from insect length in mm

(define insect-energy (lambda (insect-length)
    (* 24 (insect-mass insect-length)) ))
    ; in joules from insect length in mm

(define insect-net-energy (lambda (insect-length)
    (let* ( (assimilatable-fraction 0.75)
            (specific-dynamic-action 0.25)
            (digestive-throughput
                        (- 1 specific-dynamic-action)) )
        (* digestive-throughput assimilatable-fraction
           (insect-energy insect-length)) ) ))
    ; in joules from insect length in mm
```

This formula predicts the optimal foraging cutoff distance for an energy-maximizing forager. To explore its meaning, let's consider realistic values for the parameters, and see what optimal cutoff distances are predicted for lizards of various sizes.

1.3.2 Empirical parameters

The coefficients in the foraging strategy model pertain both to lizards and to their insect prey. Most of the physiological and ecological data to estimate these coefficients have been compiled previously [229] and are offered here in the form of Scheme functions that can be incorporated directly into Scheme programs.

Insects

Table 1.8 presents empirical parameters for insects. One begins with a measurement of an insect's length in millimeters. Its mass in milligrams is computed from its length [278, 309], and its energy content in joules is then computed from its mass [269, 8]. Finally, the net energy content in Table 1.8 is an estimate of what a lizard actually gains after digestive costs are removed [167, 274, 246, 229].

Lizards

Tables 1.9 and 1.10 present empirical parameters for lizards. In Table 1.9, one begins with a lizard's "snout-vent length," or SVL, which is the distance along its ventral surface between the tip of its nose and its vent (cloaca). The SVL omits the tail because it often detaches and then regenerates. A lizard's mass is computed from its SVL using a power law relation where the coefficient is 3.25×10^{-5} and the exponent is 2.99 [254]. For comparison, the coefficient is 4.8 ×

Table 1.9 Lizard Characteristics, Part I.

```
(define liz-mass (lambda (liz-svl)
    (* 3.25e-5 (expt liz-svl 2.98)) ))
    ; fresh weight in g from snout-vent length in mm

(define liz-sprint-speed (lambda (liz-svl)
    (* 0.75 (expt (liz-mass liz-svl) 0.35)) ))
    ; in m/s at 30 deg. C from snout-vent length in mm

(define liz-resting-met-rate (lambda (liz-svl)
    (let* ( (joules-per-cal 4.18605)
            (cal-per-ml-o2 4.75) )
          (* 0.24 (expt (liz-mass liz-svl) 0.83)
             cal-per-ml-o2 joules-per-cal
             (/ 1 60) (/ 1 60)) ) ))
    ; in joules/s from snout-vent length in mm
```

10^{-5} and exponent is 2.73 for *Anolis limifrons* of Panama [6]; the coefficient is 2.33×10^{-5} and the exponent is 3.01 for *Anolis aeneus* of Grenada based on 547 specimens ranging from 20 mm to 70 mm [347]; and the coefficient is 3.5×10^{-5} with an exponent of 3.01 for the American western fencepost lizard, *Sceloporus occidentalis* [80]. The lizard's sprint speed is then computed from its mass [152]. More recently available data specifically on the sprint speed of anoles appear in [192, 191, 196] and show that anoles can sprint about 25% faster than depicted in Figure 1.4. Next, the resting metabolic rate is given as a function of body mass as computed from the SVL [23, 24]. Moreover, the resting metabolic rate is generally measured with fasted animals resting in the dark and approximates the minimal resting costs. In the field the resting metabolic rate is higher by a factor of approximately 1.5 [64, 20, 229], and this factor appears in the first function in Table 1.10. Moving to active states, the cost of pursuing a prey item is attributed both to the incremental aerobic metabolism as measured in the respiration rate of a lizard running on a treadmill [23] and to anaerobic metabolism wherein lactic acid is reconverted to glycogen [26, 25, 23, 229]. Finally, the energetic inputs and outputs of a lizard expressed in units of joules per second may be translated into lizard eggs per 12 hours of continuous foraging, assuming an egg from *A. gingivinus* of St. Martin [19, 229], using the last function of Table 1.10.

Model coefficients

The five coefficients of the optimal foraging model (Equations 1.13, 1.16) are a_1, e_1, v, e_w, and e_p. These can all be determined from data on insect catch, insect length, and lizard SVL, as shown in Table 1.11.

The insect abundance coefficient, a_1, is estimated from insect catch expressed in units of number of insects caught per m^2 per 12 h. In practice, several (say 25) paper plates, each coated in the center with a sticky substance (Tree Tanglefoot,

Table 1.10 Lizard Characteristics, Part II.

```
(define liz-field-resting-met-rate (lambda (liz-svl)
    (* 1.5 (liz-resting-met-rate liz-svl)) ))
    ; multiplicative factor

(define liz-max-aerobic-met-rate (lambda (liz-svl)
    (let* ( (joules-per-cal 4.18605)
            (cal-per-ml-o2 4.75) )
        (* 1.96 (expt (liz-mass liz-svl) 0.76)
           cal-per-ml-o2 joules-per-cal
           (/ 1 60) (/ 1 60)) ) ))
    ; in joules/s from snout-vent length in mm

(define liz-max-anaerobic-scope (lambda (liz-svl)
    (let* ( (joules-per-cal 4.18605)
            (cal-per-ml-o2 4.75)
            (um-atp-per-mg-lactate 16.7)
            (ml-o2-per-um-atp (/ 1 290)) )
        (* (* 0.8 (liz-mass liz-svl)) (/ 1 30)
           um-atp-per-mg-lactate ml-o2-per-um-atp
           cal-per-ml-o2 joules-per-cal) ) ))
    ; in joules/s from snout-vent length in mm

(define liz-eggs-per-day (lambda (joules-per-sec)
    (let* ( (joules-per-cal 4.18605)
            (cal-per-egg-dry-wt 6160)
            (egg-dry-wt 0.13) ; for A. gingivinus
            (joules-per-egg (* egg-dry-wt
                cal-per-egg-dry-wt joules-per-cal)) )
        (* (/ joules-per-sec joules-per-egg) 60 60 12)) ))
    ; in eggs from energy content in joules
```

Table 1.11 Model Parameters

```
(define a-1 (lambda (insect-abundance-per-sq-m-per-12hr)
    (/ insect-abundance-per-sq-m-per-12hr (* 12 60 60)) ))

(define e-1 (lambda (insect-length)
    (insect-net-energy insect-length) ))

(define v (lambda (liz-svl)
    (liz-sprint-speed liz-svl) ))

(define e-w (lambda (liz-svl)
    (liz-field-resting-met-rate liz-svl) ))

(define e-p (lambda (liz-svl)
    (+ (liz-max-aerobic-met-rate liz-svl)
    (liz-max-anaerobic-scope liz-svl) ) ))
```

Tanglefoot Company, Grand Rapids, Michigan), are left in the habitat for 24 h.

The total area covered by the Tanglefoot material is determined, and the total number of insects caught is noted. This catch is then expressed as number of insects caught per m^2 of sticky surface per 12 h. The Scheme function for a-1 in Table 1.11 converts the catch measured in this way to the units used in the model and is illustrated in Figure 1.2.

The insect energy content coefficient, e_1, is estimated from insect length by the Scheme function e-1, as illustrated in Figure 1.3.

The remaining model coefficients pertain to the lizard, and are estimated from the lizard's SVL by the Scheme functions v, e-w, and e-p in Table 1.11. Figures 1.4 to 1.6 illustrate v, e_w, and e_p, respectively as a function of lizard SVL. Finally, Figure 1.7 illustrates the conversion of joules per second into lizard eggs per day.

So, now we are poised to exhibit the predicted optimal cutoff distance, r_1, and the net energetic yield, E/T, for a lizard that is obeying the optimal cutoff distance for various combinations of insect catch, insect length, and lizard SVL.

1.3.3 Optimal foraging predictions

Values for insect catch in the Lesser Antilles usually range from about 100 to 500 or more insects per m^2 per 12 h, based on data presented in more detail later. Therefore, the optimal foraging predictions are graphed for catches of 120 and 480 insects per m^2 per 12 h.

Values for average insect length range from about 1 mm in sweep samples of natural populations of arthropods from lowland tropical forests [313] to 2.5 mm in catchs from sticky plates, as also presented in more detail later. So, the optimal foraging predictions are graphed for insect lengths of 1.0, 1.5, 2.0, 2.5, and 3.0 mm.

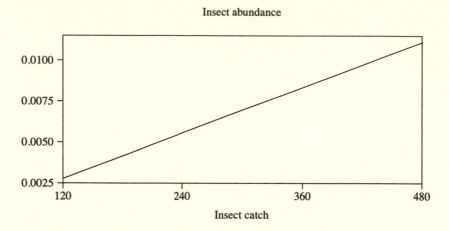

Fig. 1.2 Prey abundance, a_1, expressed as insects appearing per m^2 per s, as a function of the number of insects trapped per m^2 per 12 h.

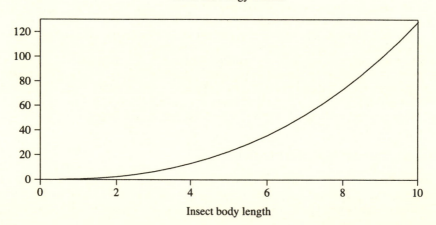

Fig. 1.3 The net energy content of an insect, e_1, in joules as a function of its body length in mm.

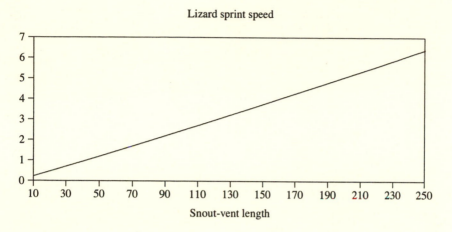

Fig. 1.4 The maximum speed of a lizard, v, in m/s at 30°C as a function of its snout-vent length in mm.

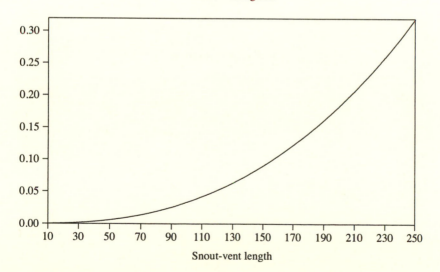

Fig. 1.5 The energy expended by a lizard while waiting for prey to appear, e_w, in joules/s as a function of its snout-vent length in mm.

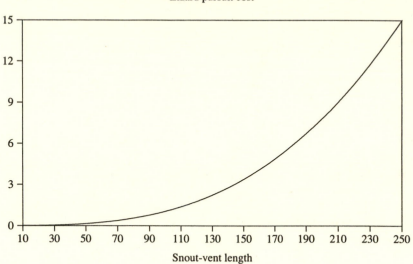

Fig. 1.6 The energy expended by a lizard while pursuing prey, e_p, in joules/s as a function of its snout-vent length in mm.

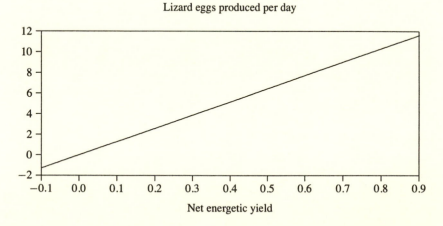

Fig. 1.7 Maximum number of lizard eggs produced per 12 h as a function of a lizard's net energetic yield in joules/s.

The snout-vent length (SVL) for anoles in the eastern Caribbean range from about 15 mm for hatchlings to about 130 mm for the largest males from the two-species islands and from Marie Galante. Therefore, foraging predictions for SVL's from 10 to 150 are presented.

Cutoff radius

Figure 1.8 illustrates the optimal cutoff distance, r_1, from Equation 1.16. The optimal cutoff distance for a given lizard SVL decreases with insect abundance and increases with insect size. For a given insect abundance and size, however, the dependence of r_1 on lizard SVL is not monotonic: Lizards of intermediate size have the largest cutoff distance.

Foraging yield

Figure 1.9 illustrates the net yield that a lizard is predicted to achieve when it does obey the optimal cutoff distance; that is, when it does pursue every insect out to a distance of r_1 and does ignore every insect beyond r_1. The net yield for these figures is obtained by evaluating Equation 1.13, the equation for energy per time, E/T, with r_1 equal to the optimal value given in Equation 1.16. As expected, the yield increases with both insect abundance and size. Of special importance, though, is the discovery of an *optimal body size* from energetic considerations alone. The energetically optimal body size (size at which the yield curves peak) varies from about 20 to 70 mm, depending on the insect characteristics. The energetically optimal body size increases markedly with the prey size and increases slightly with prey abundance.

Figure 1.9 also shows an upper limit to the body size that depends on insect characteristics. Note where the yield curves cross the horizontal axis.

The energetically optimal body sizes (SVL) predicted in Figure 1.9 are not, of course, the evolutionary optimum body sizes. But as seen later, these predictions offer a starting place. To formulate an hypothesis for the evolutionary optimum body size, growth, fecundity, and mortality throughout a lizard's life also have to be taken into account. That task is hereby placed on the agenda for later in this chapter.

Moreover, the theoretical predictions of Figures 1.8 to 1.9 come from optimal foraging theory and, as such, are static. Can a lizard learn to forage optimally according to the criterion of maximizing net yield, E/T? We now test our lizard's intelligence once more.

1.3.4 Learning to maximize yield

Our astute lizard can learn how to forage in a way that maximizes E/T, the net energetic yield. Indeed, our lizard can learn this every bit as fast as it learned how to minimize T/I, as previously illustrated.

The diskette includes a program, `learnet.scm`, that generalizes `learn.scm` to maximizing E/T. Also, `learnet.scm` contains procedures that display the data used to generate Figures 1.2 to 1.9. Once `learnet.scm` is loaded in the Scheme interpreter, one need only type (`show-all`) to obtain a table of the parameter values and numerical predictions discussed in the previous sections.

Only a few changes were needed to convert `learn.scm` to `learnet.scm` to model learning using the criterion of maximizing E/T:

Optimal home range radius

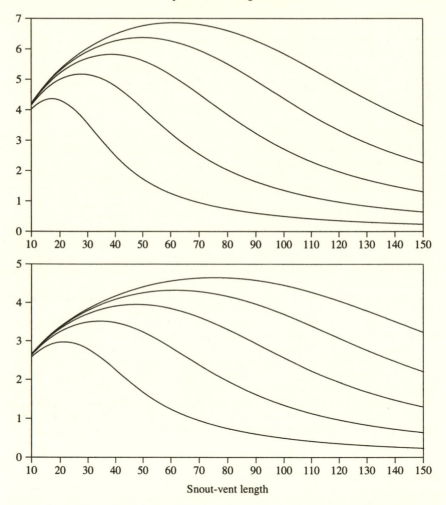

Fig. 1.8 Optimal home range cutoff radius, r_1, in m as a function of a lizard's snout-vent length in mm. Top: Prey abundance is 120 insects appearing per m^2 per 12 h; Bottom: prey abundance is 480 insects/m^2/12 h. Within each figure the curves, top to bottom, refer to insect lengths of 3.0, 2.5, 2.0, 1.5, and 1.0 mm, respectively.

Net energetic yield from optimal home range

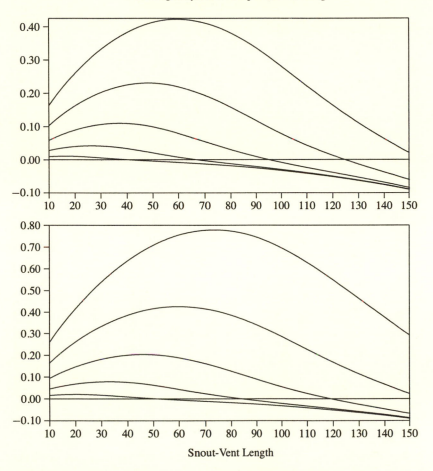

Fig. 1.9 Net energetic yield for a lizard with an optimally sized home range, E/T, in joules per second, as a function of a lizard's snout-vent length in mm. Top: Prey abundance is 120 insects appearing per m^2 per 12 h; Bottom: Prey abundance is 480 insects/m^2/12 h. Within each figure the curves, top to bottom, refer to insect lengths of 3.0, 2.5, 2.0, 1.5, and 1.0, respectively.

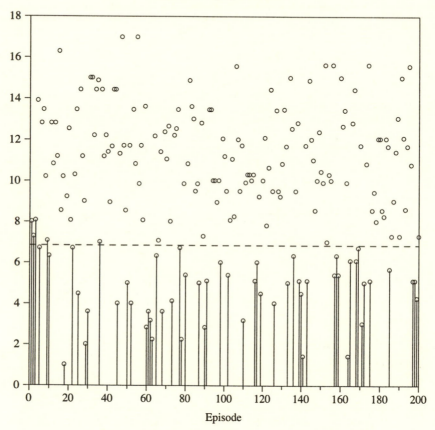

Distance of prey from lizard

Fig. 1.10 Distance in meters of prey from lizard in consecutive prey sightings. As in Figure 1.1, vertical rules indicate prey that are pursued, the horizontal dashed line indicates the optimal cutoff distance, and the lizard is shown to learn the optimal cutoff by 60 episodes. Here the lizard maximizes the net energetic yield, E/T, unlike Figure 1.1 in which the lizard minimized time per item, T/I. The lizard snout-vent length is 65 mm, the insect size is 3 mm, and the insect abundance is 120 insects per m^2 per 12 h.

- The look function now returns the insect's size, together with the waiting time and x and y coordinates as previously.

- The think function now returns a list of four components: the prey distance, as before; an estimate of the prey energy content; and the procedures to pursue and to ignore the prey.

- A result now consists of seven cumulative statistics, rather than four as before. New statistics are total energy spent in pursuit, total energy spent waiting, and total energy captured.

- The procedures to pursue and to ignore the prey developed by think now update these seven statistics, rather than only four.

- The decide function is altered to compute the energy per time expected if the prey is pursued and if ignored, and to base its recommendation on maximizing this criterion rather than on minimizing energy per item as before.

The rule of thumb for maximizing E/T is a straightforward extension of the rule for minimizing T/I. At each appearance of an insect, the rule is to do whichever action leads to a higher projected E/T. That is, when an insect appears, the projected E/T if the item is pursued is

$$(E/T)_p = \frac{E + e_1 - t_w e_w - t_p e_p}{T + t_w + t_p} \tag{1.17}$$

and the projected E/T if the item is ignored is

$$(E/T)_i = \frac{E - t_w e_w}{T + t_w} \tag{1.18}$$

where t_p is the anticipated time to pursue the insect and to return to the perch, t_w is the time since the last insect appeared, e_1 is the energy content of a prey item, e_p is the energy expended during pursuit and return to the perch, e_w is the energy expended while waiting, T is the total elapsed time when the lizard finished with the previous insect, and E is the total energy captured as of that time. The rule is to do whichever action leads to

$$\max[(E/T)_p, (E/T)_i] \tag{1.19}$$

To use the computer program with a specific example, the lizard's SVL is stipulated, along with the insect size and abundance. Consider a 65-mm lizard and 3-mm prey whose abundance is 120 insects per m^2 per 12 h. From Figure 1.8, this situation implies an optimal cutoff radius of 6.8 m. These particular values are chosen here so that the results can be graphed using the same scale as Figure 1.1, facilitating visual comparison. Essentially the same results occur with other lizard and insect parameters.

As before, to run the program type

```
(define experience (live 200 birth))
(recall experience)
```

Figure 1.10 shows the sentient lizard learning the optimal cutoff radius within 60 episodes—the same remarkable quickness of mind previously demonstrated. By 60 episodes the lizard does not chase prey beyond the dashed line, and it does not ignore prey within the dashed line—that is, it forages optimally. After 100 episodes, 296 seconds (nearly 5 minutes) have elapsed: Of the 100 insects sighted, 29 were captured and 71 were ignored, for a cumulative yield of 127 joules and for an E/T of 0.42 joules/s. This E/T corresponds with the prediction of Figure 1.9.

So, once again we have discovered a zoologically plausible behavioral rule for the lizard to follow. Following this behavioral rule of thumb leads the lizard to forage optimally. In a nutshell, here's how the illustration worked. The first prey was sighted 8.0 m away, and by assumption the lizard chased the first prey it saw regardless of where it was. Its E/T was then set initially at 0.28. The second prey was sighted at 7.2 m. The lizard evaluated the two possibilities: ignoring the prey versus chasing the prey, and it did whichever of these would produce the higher E/T. In fact, the lizard chased the prey, and its E/T rose to 0.31. Similarly, it chased the third prey, which was sighted at 8.1 m, and its E/T dropped slightly to 0.30. However, its E/T would have dropped even more had the prey been ignored. Then the lizard ignored its fourth sighting at 13.4 m, and its E/T dropped still further to 0.29. Here its E/T would have dropped even more had the lizard chased the insect. Then the lizard chased its fifth prey, sighted at 6.7 m, and E/T rose back to 0.30. Thus, by asking at each prey sighting how chasing or ignoring the prey would affect its next cumulative E/T, and acting accordingly, the lizard eventually attained an E/T that faithfully described the maximal yield possible in the environment. As the lizard's realized E/T approached this maximal yield, its discrimination of the optimal cutoff distance became more accurate. After 100 prey sightings the lizard's realized E/T was within two decimal places of the theoretical maximum yield, and its decision of where to set the cutoff distance almost exactly matched the theoretical optimum cutoff distance.

While the sentient lizard has learned to forage optimally from the standpoint of maximizing E/T, this optimum is not typically the "very best" behavior that one can imagine. The lizard learns a short-sighted optimum, one that does not take population-dynamic feedbacks into account. The lizard does not consider the effect, if any, of its foraging on the prey's renewability and therefore does not generally harvest for a maximum sustainable yield of insect prey.

1.3.5 Insects and diets, St. Eustatius

By now you've probably built up an appetite for some data. How does our astute, yet theoretical, lizard compare with the real thing?

St. Eustatius in the northeastern Caribbean is one of the Netherlands Antilles. It belongs to the St. Kitts bank and shares the same species of anoles with St. Kitts and Nevis to its south. St. Eustatius is xeric with a small human population; it has proved a good location for experimental field studies [240, 298]. Two species of anoles occur on St. Eustatius: the small brownish *A. schwartzi* and the larger greenish *A. bimaculatus*.

Body size

Figure 1.11 illustrates a typical distribution of body sizes in the St. Eustatius species. The distributions include all ages and both sexes. These specimens were collected

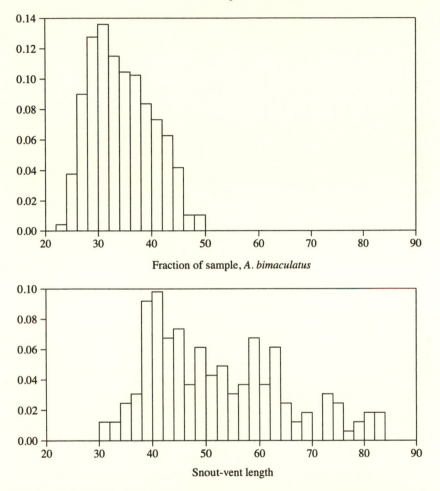

Fig. 1.11 Distribution of snout-vent length in mm from samples collected on St. Eustatius during February 1982. Top: *A. schwartzi*, sample size is 480 lizards (both sexes and juveniles combined), mean SVL is 34.1 mm, and standard deviation is 5.8 mm. Bottom: *A. bimaculatus*, sample size is 160 lizards (both sexes and juveniles combined), mean SVL is 51.6 mm, and standard deviation is 12.5 mm.

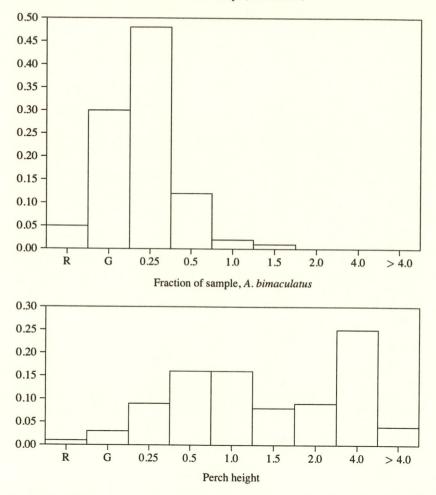

Fig. 1.12 Distribution of perch heights on St. Eustatius during March, April, and May 1982 ("R" for rocks, "G" for ground). Top: *A. schwartzi*, sample size is 542 observations (both sexes and juveniles combined), mean perch height is 0.14 m, and standard deviation is 0.19 m. Bottom: *A. bimaculatus*, sample size is 553 observations (both sexes and juveniles combined), mean perch height is 1.83 m, and standard deviation is 1.59 m.

Table 1.12 Abundance of *Anolis* Lizards, St. Eustatius

Dates	Species	Procedure	N/100 m^2(SE)
Jan. 16–18 1983	A. schwartzi	3-day independ.	64.0(1.1)
	A. bimaculatus	3-day independ.	30.0(3.6)
June 23–24, 1983	A. schwartzi	2-day independ.	47.8(2.0)
	A. bimaculatus	2-day independ.	22.2(3.6)
Mar. 28–30, 1984	A. schwartzi	3-day independ.	98.5(2.1)
	A. bimaculatus	3-day independ.	25.6(2.2)
July 26–28, 1984	A. schwartzi	3-day $u_{23} \neq 0$	56.1(1.0)
	A. bimaculatus	3-day $u_{23} \neq 0$	26.7(5.4)

from the wild more or less at random during February 1982 and were introduced to experimental exclosures for studies of interspecific competition [298], as discussed further in the next chapter. No competition between these species was found. Adult males are larger than adult females, and the upper part of the distributions are males. Note the largest male of *A. bimaculatus* is about 85 mm, compared with about 50 mm for *A. schwartzi*.

Perch heights

The St. Eustatius species also differ in where they characteristically perch in the vegetation. Figure 1.12 illustrates the distribution of perch heights [297]. The larger *A. bimaculatus* generally perches above the smaller *A. schwartzi*. In practice, one tends to locate *A. bimaculatus* individuals by looking up, often into the canopy, and individuals of *A. schwartzi* by looking down, often on leaf litter and rocks.

Our studies have shown that body size and perch height are the only major differences between these species. On other islands, however, the species may also differ significantly in whether they perch on tree trunks versus thin branches, twigs, or bushes, and whether they perch in sunny versus shady microclimates. These other possible dimensions of separation, perch diameter and perch micro-climate, are not evident in the northern Lesser Antillean communities, including St. Eustatius, a point also discussed further in the next chapter.

Abundance

Table 1.12 presents data on the abundance of anoles on St. Eustatius in units of number individuals per 100 m^2. (For census methods, cf. [141].) These data, not published previously, were obtained in a 160.1 m^2 quadrat used as an undisturbed control for the experimental studies alluded to before. The quadrat was located along the side of a large ravine that opens toward the northeast and is on the northeast side of the island. The quadrat had a ground slope of about 18° facing northwest (about 320°), and the elevation was about 110 m above sea level. The site was wooded, with the canopy height at about 6.5 m and with an understory at about 3 m. About 10% of the ground was rocks, with the remainder being leaf litter and soil. There were about 10 trees (stems > 15 cm in diameter) per 100 m^2.

The table shows that the species of smaller lizards is two to four times more abundant than the species of larger lizards, depending on time of year. The total number of lizards varies from about 70 to 120 per 100 m^2, which accords with the rough generalization that West Indian anole abundance is about one lizard per m^2. The abundance varies seasonally, with the high in the winter to early spring soon after the rainy season, and the low in the summer during the dry season. Further data on abundance appear in the next chapter.

Territory size

The abundance data, together with the knowledge that anoles of all ages and both sexes are territorial, imply that the horizontal projection of the home range area is approximately the reciprocal of the population density. That is, the home range area of the smaller species, *A. schwartzi*, varies from about to 1 to 2 m^2, and that of the larger species, *A. bimaculatus* from about 3 to 4 m^2. Furthermore, unpublished measurements of the home ranges of individual *A. pogus*, the ecological equivalent on St. Martin of *A. schwartzi*, also directly show a home range area of about 1 m^2 at a site similar to that on St. Eustatius [28].

Similarly, the home ranges of adult female *Anolis aeneus* of Grenada have been reported between about 2 m^2 to 4 m^2 in scrub habitat [347]. Stamps offers a log-log plot of home range area in m^2 and lizard mass in g that can be fitted roughly with a straight line. Although there is great variability between sexes and across habitats, the data are approximated by the relation that home range area in square meters equals 1.3 times body mass in grams to the 1.1 power; or equivalently, that home range area in m^2 equals 3×10^{-5} times SVL in mm to the 3.33 power. These relations probably can be used for ballpark guesses of territory sizes for small Caribbean anoles such as *A. aeneus* that range from 20 to 70 mm in snout-vent length. Of course, territory area should not be thought of as a physiological parameter even though it may vary with body dimension approximately as a power law, because territory size is the net result of both physiology and local ecological conditions including population density. Diagrams and further studies of territories for juveniles, females, and males of *A. aeneus* in several habitats appear in [343, 353, 347, 351, 352, 354], and a review of *Anolis* home ranges appears in [318].

Foraging radius

A guess at the foraging cutoff radius can be derived from the estimates of home range area. If we suppose the home range is a semicircle of radius, r_1, then its area, A_h, is $(1/2)\pi r_1^2$. So, we may guess that $r_1 = \sqrt{2A_h/\pi}$. Hence r_1 for *A. schwartzi* appears to be about 0.8 to 1.1 m, and for *A. bimaculatus* about 1.3 to 1.6 m.

These guesses at r_1 should, of course, be viewed with caution. The reciprocal of the population density provides only the projection of the home range area on the ground. In typical habitat an anole may travel a long distance vertically while moving only slightly in a horizontal dimension. Indeed, the most comparable species from Puerto Rico, the ubiquitous *A. stratulus*—a 40 mm arboreal anole

Table 1.13 *A. schwartzi*, St. Eustatius, stomach contents of 80 lizards, classified to Order.

Taxon	Mean No.	Mean Vol.	% No.	% Vol.
Annelida	0	0	0	0
Gastropoda	0.03	0.020	0.12	0.11
Acarina	0.15	0.002	0.71	0.01
Araneida	0.21	1.200	1.00	6.52
Chilopoda	0	0	0	0
Diplopoda	0	0	0	0
Isopoda	0.21	0.319	1.00	1.73
Scorpionida	0.01	+	0.06	+
Coleoptera, adults	0.65	1.589	3.06	8.63
Coleoptera, larvae	0.34	0.168	1.59	0.91
Collembola	0.11	0.010	0.53	0.06
Diptera, adults	2.88	4.064	13.55	22.08
Diptera, larvae	0.15	0.047	0.71	0.26
Hemiptera	0.01	0.007	0.06	0.04
Homoptera	2.49	1.331	11.73	7.23
Hymenoptera, formicidae	11.73	4.454	55.27	24.20
Hymenoptera, other	0.36	0.292	1.71	1.59
Isoptera	0.01	0.008	0.06	0.05
Lepidoptera, adults	0.28	1.417	1.30	7.70
Lepidoptera, larvae	1.14	1.824	5.36	9.91
Orthoptera	0.06	0.681	0.29	3.70
Thysanoptera	0.13	0.004	0.59	0.92
Other insects, adults	0.20	0.638	0.94	3.47
Other insects, larvae	0.08	0.332	0.35	1.80

that in the rain forest of the Luquillo Mountains lives from ground level to the canopy at 20 m—has a home range 6 m in diameter [267]. Thus, we will take r_1 as being about 3/4 to 2 m for average prey, but it could be as large as 5 m for rare and particularly desirable prey.

The most comparable data on moving distances and time intervals between moves comes from Hispaniola [224, Figs. 2 and 3]. The distribution of waiting times between moves for seven species are all exponential in shape with the maximum time between moves being up to 30 minutes, and the maximum distances of the moves extending to 60 cm. Most of the moves are reported to be foraging strikes, but they include moves for other purposes as well. The average time between moves varies from about 3 to 6 minutes, and the average distance per move varies from about 10 to 30 cm, depending on the species.

Diet

Tables 1.13 and 1.14 show what *A. schwartzi* and *A. bimaculatus* harvest from their home ranges [297]. Notice the wide spectrum of prey consumed by both

Table 1.14 *A. bimaculatus*, St. Eustatius, stomach contents of 50 lizards classified to Order

Taxon	Mean No.	Mean Vol.	% No.	% Vol.
Annelida	0.08	0.013	0.31	0.02
Gastropoda	0.08	0.482	0.31	0.70
Acarina	0.02	0.003	0.08	+
Araneida	0.10	0.129	0.39	0.19
Chilopoda	0.02	2.532	0.08	3.67
Diplopoda	0.02	0.018	0.08	0.03
Isopoda	0.04	0.026	0.16	0.04
Scorpionida	0	0	0	0
Coleoptera, adults	1.48	14.143	5.77	20.47
Coleoptera, larvae	0.02	0.003	0.08	+
Collembola	0.02	0.002	0.08	+
Diptera, adults	0.56	4.291	2.18	6.21
Diptera, larvae	0.20	0.214	0.78	0.31
Hemiptera	0.02	0.003	0.08	+
Homoptera	1.44	0.772	5.62	1.12
Hymenoptera, formicidae	8.52	4.509	33.23	6.53
Hymenoptera, other	2.12	3.402	8.27	4.92
Isoptera	0.54	1.779	2.11	2.58
Lepidoptera, adults	0.36	1.989	1.40	2.88
Lepidoptera, larvae	9.58	23.129	37.36	33.48
Orthoptera	0.12	9.619	0.47	13.92
Thysanoptera	0.02	+	0.08	+
Other insects, adults	0.16	1.686	0.62	2.44
Other insects, larvae	0.12	0.339	0.47	0.49

Table 1.15 Arthropods caught by lizards on St. Eustatius in May 1982. (Liz. is number of lizards, Arth./Liz. is number of arthropods per lizard, standard deviation in parentheses.)

Species	Liz.	SVL	Arth./Liz.	Length	% in Length Class		
					1–3	4–9	≥ 10
A. schwartzi	247	40.5	18.4(12.7)	2.3(1.5)	86.7	12.7	0.6
A. bimaculatus	87	56.3	19.6(20.5)	3.0(1.9)	71.9	27.2	0.9

species, spanning at least 18 orders of insects. This observation prompts another approximate rule: Anoles are generalized predators of squiggley things. Indeed, one can actually catch an anole by dangling a fishing rod with a line and artificial fly.

Although both anole species consume the same general kinds of prey, a comparison of Tables 1.13 and 1.14 shows quantitative differences between the species. *A. schwartzi* eat lots of ants; 55% of its prey are ants, and its diet also includes 13% fruit flies, and 12% leaf hoppers (and other homoptera). These are relatively small insects. In contrast, ants comprise only 33% of the diet of *A. bimaculatus*. Indeed, at 37% caterpillars are slightly more numerous in the diet of *A. bimaculatus* than ants; and beetles and crickets also make significant contributions. These insects are larger than the ants, fruit flies, and leaf hoppers that predominate in the diet of *A. schwartzi*. Indeed, Table 1.15 indicates that the average prey size in the stomachs of *A. schwartzi* is 2.3 mm, and in stomachs of *A. bimaculatus* is 3.0 mm [298]. The table also shows the distribution of the stomach contents into length classes. The insect use of *A. bimaculatus* is shifted into larger insect lengths relative to that of *A. schwartzi*.

Incidentally, the average snout-vent length of the specimens of Table 1.15 is, for both species, about 5 mm larger than the averages given in the histograms of Figure 1.13. This difference primarily represents growth of the specimens between February and May 1982.

Similar data on dietary differences between coexisting anole species in the eastern Caribbean have been presented for Grenada [312, 356, 355], and recently diet and foraging data for the solitary anole on Dominica have been determined [50].

Selectivity

Table 1.16 shows the distribution of insects in the habitat, as assayed by the catches on paper plates, covered in their centers with sticky Tanglefoot, that were placed on the forest floor [298]. Comparing Table 1.16 with 1.15 reveals that both species do not simply consume prey in the proportions encountered. Both species have more large prey in their stomachs relative to the proportions available in the environment. But *A. bimaculatus* is more selective of large prey than *A. schwartzi* is, and has larger prey in its stomach as a result.

The greater selectivity for large prey by *A. bimaculatus* relative to *A. schwartzi* probably underlies the rather slight taxonomic differences in their diets. The greater selectivity for larger prey leads *A. bimaculatus* to consume more of the insect taxa that are typically large, such as caterpillars and crickets, as noted above.

Table 1.16 Arthropods caught in traps on St. Eustatius in April, May 1982. 1982. (Abund. is number of arthropods per m² per 12 h, standard deviation in parentheses.)

Date	Catch/Trap	Traps	Length	% in Length Class			Abund.
				1–3	4–9	≥ 10	
Apr 15	57.5(101.6)	40	2.2(1.2)	96.1	3.5	0.4	915
May 24	17.7(13.3)	40	2.2(1.5)	97.3	1.4	1.3	281
May 31	25.9(30.1)	120	2.3(1.1)	95.0	4.8	0.2	413

Furthermore, *A. bimaculatus* may perch higher than *A. schwartzi* to obtain a higher vantage point from which to search for large and relatively rare prey. But the different perch heights then, in turn, can also partly account for taxonomic dietary differences. *A. schwartzi*, for example, is closer to fruit flies on the forest floor than is *A. bimaculatus*. Conversely, *A. bimaculatus* is closer to caterpillars on leaves than is *A. schwartzi*. Both species are probably equally close to ants, however, as ant trails are readily seen on the ground, on tree trunks, and in the canopy.

Still, much of *A. bimaculatus* foraging activity is on the ground, and *A. bimaculatus* individuals are often seen facing the ground even though they are typically perched about two meters above the forest floor. The presence of beetles and crickets in the diet of *A. bimaculatus* indicates a willingness to chase large ground-dwelling prey from their relatively high perches.

Thus, the body size difference between *A. bimaculatus* and *A. schwartzi* appears to explain why they consume different prey sizes, why they perch at different heights in the vegetation, and why they have some differences in the taxonomic composition of their diets.

Comparison with model

With this information we can reflect on how well the preceding theory accounts for *Anolis* foraging. Does our theoretical lizard look like a real *Anolis*? I think it is beginning to; it is starting to resemble a real lizard, not just a sci-fi creation, though our theoretical lizard still has some way to go before it is an accurate representation of an anole.

The snout-vent lengths of real anoles nicely span the range of body sizes that produce a high energetic yield, E/T, according to Figure 1.9. According to the figure, anoles from 15 mm to 150 mm are energetically viable, and anoles of these sizes in fact occur. The best size from purely energetic considerations is 60 to 70 mm for many parameter combinations. These are typical sizes for anoles, although as will be seen later, the optimal size from an evolutionary standpoint will be different, indeed somewhat smaller, once survivorship is taken into account.

However, the predicted yields are high when converted to units of eggs produced per day. For example, a yield of 0.4 J per s corresponds to 4 eggs per 12 h of foraging. The shortest reported interval between egg laying is 11.7 days for *A. aeneus* of Grenada [344], and estimates of the interval between ovulations vary from one to two weeks [9] to two to four weeks [296].[3] Thus a yield of about 0.1

---

[3] Anoles lay one egg at a time from alternate ovaries. During peak reproduction season, females typically

to 0.05 eggs per 24 h is reasonable for an anole, and the predicted yield is off by about two orders of magnitude. The model's predictions may not be quite this far wrong though. A 12-h day is undoubtedly much more time than is available to an anole, and maintenance costs for the remaining 12 h have not been included; corrections for these considerations will be brought in later in this chapter.

There is another discrepancy too. The real home range radii of anoles (about 3/4 to 2 m) are smaller than the radii predicted in Figure 1.8. This discrepancy too is not as serious as it might seem because a horizontal projection of the home range radius underestimates the actual cutoff radius. Still, both the discrepancies between the model and real anoles, if genuine, may be interrelated. A smaller home range would lead to a lower yield, and if corrections to the model lead to a smaller predicted home range, the predicted yield would be lower too.

If we suppose that the model is inaccurate, there are two possibilities. Perhaps the theory's predicting larger home ranges and higher yields than real anoles reflects the theory's focus on a solitary animal—an animal without neighbors. Perhaps territorial squabbles with neighbors squeeze each lizard's territory into a smaller area than predicted by theory for a solitary animal. So far, our theoretical lizard is smart, a bit too fat, and lonely. One approach would be to include social interaction in the model. Alternatively, the power of a model for a solitary animal has not yet been exhausted. The prey in the model so far are assumed to remain stationary at the coordinates where they were originally sighted. Clearly, some kinds of insects have ample time to escape if a lizard is trying to run them down over a distance of 3 to 4 meters. Other insects, such as caterpillars, are almost sessile and can be captured regardless of the distance at which they are sighted. Nonetheless, allowing for prey escape is certain to lower both the predicted cutoff distance and the yield, thus ameliorating both of the model's discrepancies. Perhaps including prey escape will make our theoretical lizard smart, lean, and lonely, and more realistic.

1.3.6 Prey escape

Anoles typically wait for an insect to alight and then scamper to catch it [225]. The ability of anoles to detect motion such as a prey alighting some distance away is highly refined [115, 116]. When anoles miss, it's because the insect flies away before they arrive. Some insects seem naturally edgy, and tend to fly away regardless of circumstances, while others seem to fly off only when somehow alerted to danger. In either case, an anole may have only a short time to run from its perch to the insect's position before the insect escapes.

Flightiness

To include the possibility of prey escape into the foraging theory, it seems reasonable to suppose that a prey has an index of flightiness, say f_1 for a prey of type 1, that is related to the time it will remain stationary and vulnerable to predation. If f_1 is the probability per second that the insect flies away from the spot where it is located, then the probability of its remaining where it alighted for t or more seconds is distributed exponentially with parameter f_1. That is, the probability of an insect of type 1 remaining where it alighted for t seconds or more is $\exp(-f_1 t)$. If the place where the insect alighted is r meters away from the lizard, then the

---

have a shelled egg in one oviduct, an unshelled egg in the other, and one enlarged follicle.

time the lizard needs to reach the insect is r/v where v is the lizard's sprint speed. So, the probability that a lizard can catch the insect is

$$\text{Probability of a particular capture} = e^{-f_1 r/v} \qquad (1.20)$$

Now if r_1 is the cutoff distance, the fraction of all the pursuits within this cutoff distance that are successful, F_1, is

$$F_1(r_1) = \int_0^{r_1} \left(\frac{a_1 \pi r}{\int_0^{r_1} a_1 \pi r dr} \right) e^{-f_1 r/v} dr \qquad (1.21)$$

$$= \frac{\pi a_1 \left(v/f_1 \right) \left[v/f_1 - \left(r_1 + v/f_1 \right) e^{-r_1 f_1/v} \right]}{a_1 \pi r_1^2 / 2} \qquad (1.22)$$

The flightiness index, f_1, can be measured as the reciprocal of the average time an insect remains at the spot where it alighted in the face of an attacking lizard. An index of 0 indicates a stationary prey, that is, one that remains an infinite time at the spot where it alighted. Undisturbed caged houseflies have been measured to have an f_1 of 0.0023 per s, corresponding to an average residence time at the spot where they alighted of about 7.5 minutes [148]. This number is clearly too low to apply in field conditions when the fly is being attacked. Anyone could swat a fly in 7.5 minutes.

Small cyprinid fish begin flight upon attack by largemouth bass when the rate of change of the angle subtended the predator at the prey's eye exceeds some threshold value [90]. That is, a fast-moving distant predator evokes the same alarm as a slow-moving nearby predator. Perhaps similar behavior is exhibited by insects under attack by a lizard. At this time however, I'm simply not aware of any data on insects that allow the flightiness coefficient, f_1, to be estimated; it must be measured specifically during the time when an insect is under a lizard's attack. Therefore, to move ahead, we'll have to sacrifice a degree of freedom in testing this theory by estimating f_1 from the data with which the model is being compared.

A guess at f_1

We shall estimate f_1 from assuming that the average residence time of the insect, $1/f_1$, equals the lizard's transit time to the outer edge of its home range, r_1/v. Solving for f_1 yields

$$f_1 = v/r_1 \qquad (1.23)$$

Specifically, if we suppose a 35-mm lizard like *A. schwartzi* has the outer edge of its home range about 3/4 m away from its perch, and with a sprint speed of 0.75 m per s from Figure 1.4, then the flightiness index for its prey should be about 1.0 per s. Because the reciprocal of 1 is also 1, this f_1 implies that the lizard has about 1 second to effect its capture. Similarly, if we suppose a 55-mm lizard like *A. bimaculatus* has the outer edge of its home range about 1.5 m away from its perch, and with a sprint speed of 1.25 m per s from Figure 1.4, then f_1 works out to about 0.8 per s. This f_1 implies the lizard has about 1.2 seconds in which to effect its capture. So, these considerations suggest that realistic values of f_1 include 0.5 and 1.0.

In the future, considerations from functional morphology can be incorporated into the f_i coefficients. Although we are viewing the f_i as properties of the insect, these coefficients are more correctly viewed as properties of both the insect and the lizard. A bug can more easily escape from a clumsy lizard than from an agile lizard. E. E. Williams [383, 385] introduced the idea of an "ecomorph" to describe the characteristic shapes and sizes of anoles that occupy typical kinds of vegetation structure. Subsequent work has shown how locomotor performance in various microhabitat structures is affected by morphology [223, 198, 192]. Thus, the f_i of a given insect depends on how effectively the lizard ecomorph functions in the vegetation structure where the insect occurs. It should be noted that functional morphology is also implicit in the model's energetic parameters and in the sprint velocity.

E/T with escape

Now let's see what the impact of this prey escape is for the predictions from optimal foraging theory. The expression for E/T originally given in Equation 1.13 needs modification to incorporate prey escape. The term involving the prey energy content, e_1, should now be multiplied by $F_1(r_1)$ to indicate the fraction of chases that are successful. The terms for the time and energy involved in waiting and pursuit remain unchanged.

$$E/T = \frac{e_1 F_1(r_1) - \frac{2}{a_1 \pi r_1^2} e_w - \frac{4r_1}{3v} e_p}{\frac{2}{a_1 \pi r_1^2} + \frac{4r_1}{3v}} \tag{1.24}$$

But now the plot thickens. Setting the derivative of E/T with respect to r_1 equal to 0 yields an equation that cannot be solved explicitly, and we must proceed numerically.

A well-known technique for obtaining roots to a nonlinear equation is called the Newton-Raphson method, and it works as follows. To find the root of the equation $f(x) = 0$, we expand $f(x)$ to first order around some initial guess at where the root is, x_0. That is, $f(x) \approx (df(x_0)/dx)(x - x_0) + f(x_0) = 0$. The idea is then to rearrange this equation by solving for x, and to view x as a new and improved guess at where the root is. Call this new x as x_1. Then the process is repeated to generate still another guess, x_2, and so forth until the guesses converge to a solution. Thus, guess $n + 1$ is computed from guess n as

$$x_{n+1} = x_n - \frac{f(x_n)}{df(x_n)/dx} \tag{1.25}$$

Cutoff radius and yield

Figures 1.13 and 1.14 offer the predictions of this revised theory for the optimal foraging cutoff distance and for the net yield to an optimal forager. These figures assume an insect abundance of 480 insects per m^2 per 12 h. They may be compared with the lower panels in Figures 1.8 and 1.9 for stationary prey. In these figures, the top panels refer to insect flightiness coefficients of 0.5, corresponding to a 2-second period during which the lizard must reach its prey, and the bottom panels to flightiness coefficients of 1.0, corresponding to a 1-second chase. The data

Optimal home range radius

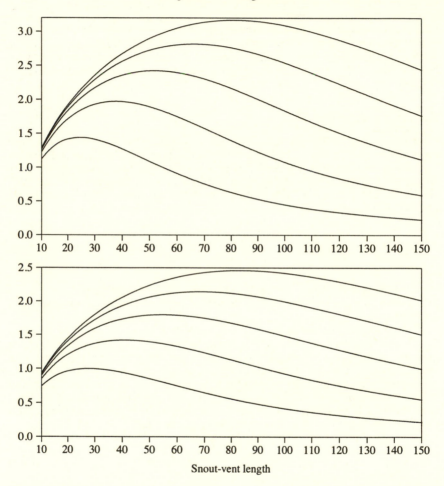

Fig. 1.13 Optimal home range cutoff radius, r_1, in m as a function of a lizard's snout-vent length in mm. Prey abundance is 480 insects appearing per m^2 per 12 h. Top: Insect flightiness coefficient is 0.5; Bottom: Insect flightiness coefficient is 1.0. Within each figure the curves, top to bottom, refer to insect lengths of 3.0, 2.5, 2.0, 1.5, and 1.0 mm, respectively.

for these figures were obtained from the program `forage.scm` included on the diskette by typing `(show-r1-opt)`.

The lower panels of Figures 1.13 and 1.14, in which f_1 is 1.0, show foraging cutoff distances and energetic yields that are approaching realistic values for anoles. Specifically, the foraging cutoff distances roughly accord with those estimated earlier for the St. Eustatian anoles. In itself, this agreement with the data is not impressive because the value of f_1 was originally determined from these data to bring about the agreement. Next, the energetic yield is beginning to seem plausible. With an average prey length of 2.0 mm, the best SVL is 50 mm, which returns a yield of 0.3 eggs per 12 h (cf. Figure 1.7), implying about 36 continuous hours of foraging to make an egg. With a prey size of 2.5 mm the best SVL is 70 mm for a yield of 0.9 eggs per 12 h, implying about 11 h of continuous foraging to make an egg. As mentioned earlier, a day usually does not contain a full 12 h of foraging time, and when the maintenance costs of resting at night are included, the time needed to produce an egg becomes much longer.

Diet

But now a new discrepancy emerges between the model and actual foraging behavior of *Anolis*; it concerns the total number of items caught per day. The number of items pursued during some period, say T, is found by dividing the average time expended per item into T, to indicate the number of items that can be squeezed into the time period. The period, T, will typically be 12 h (43,200 s). Because the time per item is the sum of the waiting time and pursuit time, the number of items pursued during T is

$$\frac{T}{\frac{2}{a_1 \pi r_1^2} + \frac{4r_1}{3v}} \tag{1.26}$$

To find the total number of items captured, we need to multiply the above expression by the fraction of pursuits that is successful, $F_1(r_1)$:

$$\frac{T}{2/(a_1 \pi r_1^2) + 4r_1/(3v)} \frac{\pi a_1(v/f_1) \left[v/f_1 - \left(r_1 + v/f_1 \right) e^{-r_1 f_1/v} \right]}{a_1 \pi r_1^2/2} \tag{1.27}$$

This formula works out to 853 insects per 12 h of continuous optimal foraging for a 50-mm lizard with 2-mm prey having an abundance of 480 insects per m^2 per 12 h. This daily catch amounts to slightly more than one insect caught per minute, an estimate off by an order of magnitude or more. A lizard probably catches an item no more often than every 10 minutes, and Table 1.15 indicates an average stomach content of about 20 insects per lizard. Hence, a reasonable guess is that an anole catches about 30 to 50 insects per day, because observations on the Puerto Rican anole *A. stratulus* show that the number of prey caught per day is approximately twice the number found per individual in stomach contents [190, 266]. Therefore, the theory as developed so far is still seriously inaccurate.

The problem with the theory at this stage is its focus on just one prey size, an "average insect size." Yet, for an anole, catching one cricket is worth hundreds of ants. By using multiple prey types in the theory, the consumption of ants will be greatly lessened once the possibility of catching crickets is included. Therefore, we will soon move on to introduce the last major enhancement of the theory, provision for multiple prey types.

Net energetic yield from optimal home range

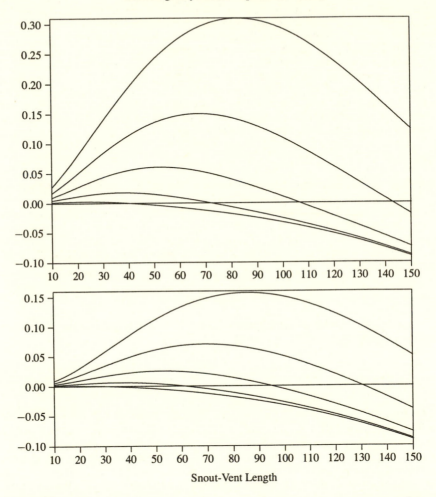

Fig. 1.14 Net energetic yield for a lizard with an optimally sized home range, E/T, in joules per second, as a function of a lizard's snout-vent length in mm. Prey abundance is 480 insects appearing per m^2 per 12 h. Top: Insect flightiness coefficient is 0.5; Bottom: Insect flightiness coefficient is 1.0. Within each figure the curves, top to bottom, refer to insect lengths of 3.0, 2.5, 2.0 1.5, and 1.0, respectively.

Learning

But before jumping in with multiple prey, an illustration is offered that the sentient lizard can learn where the optimal cutoff radius is, even though prey occasionally escape. The lizard is assumed to know how to estimate the odds of making any particular capture given the distance to the item, and to weigh its value of the prey's energetic content by these odds. As before, at each appearance of an insect, the rule of thumb is to do whichever action leads to a higher projected E/T. When an insect appears, the projected E/T if the item is pursued is

$$(E/T)_p = \frac{E + p_1 e_1 - t_w e_w - t_p e_p}{T + t_w + t_p} \tag{1.28}$$

and the projected E/T if the item is ignored is

$$(E/T)_i = \frac{E - t_w e_w}{T + t_w} \tag{1.29}$$

where p_1 is the probability that the lizard can successfully reach the insect before it escapes (Equation 1.20). The rule is to do whichever action leads to

$$\max[(E/T)_p, (E/T)_i] \tag{1.30}$$

A more elaborate simulation could explore how the lizard learns what the odds are for capturing prey. Figure 1.15 (top) shows E/T approaching its theoretically predicted value, and (bottom) shows the distance at which insects are caught. Again observe that the lizard makes mistakes only early in the foraging period. This illustration was made with using the program `forage.scm`, included on the diskette.[4]

1.3.7 Final theory with multiple prey

The anoles of St. Eustatius reveal that selectivity for large prey increases with body size: Large lizards eat large prey. An informative foraging theory should be able to predict how the entire diet of a lizard in any environment depends on its body size and other physiological parameters. And, in particular, it should predict the relation between a lizard's average prey size and its body size.

We will discuss three prey types in hopes that a lizard's prey can be reduced to three categories: small, medium, and large, corresponding roughly to ants, large flies, and crickets. If needed, the extension to even more categories could proceed along the same lines.

With three prey, the lizard has three optimal cutoff radii, r_1, r_2, and r_3, corresponding to the three prey types. For purposes of notation, we will order the prey from smallest to largest, so that r_1 refers to the cutoff radius for the smallest category of insects, and so forth. It would be nice if we could compute these three radii independently of each other simply by using Equation 1.25 three times, once for each prey category. Alas, the optimal foraging radii are interdependent and must be computed simultaneously. We seek, therefore, an optimal vector, (r_1, r_2, r_3), that maximizes E/T.

[4]The parameters are xmax = 8, ymax = 8, look-xy-step = 0.1, look-t-step = 0.25, and nbig = 1000. The run was made without any special initialization to the random number generator.

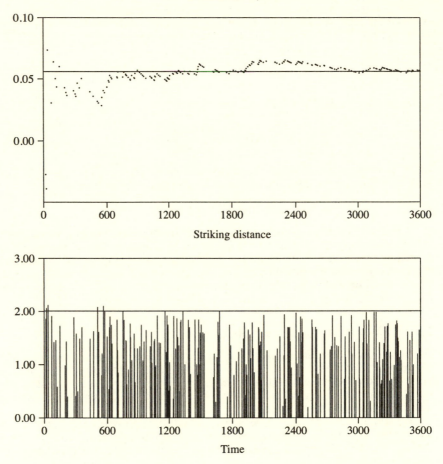

Fig. 1.15 Top: Net energy gained per elapsed time in joules per second during one h of foraging, as noted after each prey capture. Expected yield from optimal home range shown as horizontal line. Bottom: Distance in meters of successful strikes as function of time in seconds. Optimal cutoff radius shown as horizontal line. Insect length is 2.5 mm, prey flightiness coefficient is 1, abundance is 480 insects per m^2 per 12 h, lizard SVL is 45 mm, and lizard's memory extends to beginning of foraging period.

E/T with multiple prey

The form of the expression for E/T is

$$E/T = \frac{\bar{e} - t_w e_w - t_p e_p}{t_w + t_p} \tag{1.31}$$

where $\bar{e}$ is the average net energy content of an insect, t_w and t_p are the average times of waiting and pursuing insects, and e_w and e_p are the average energy expenditures during waiting and pursuit. The full expression for E/T with three prey types then is

$$E/T =$$

$$\left(\frac{a_1 e_1 r_1^2 F_1(r_1) + a_2 e_2 r_2^2 F_2(r_2) + a_3 e_3 r_3^2 F_3(r_3)}{a_1 r_1^2 + a_2 r_2^2 + a_3 r_3^2} \right.$$

$$\left. - \frac{2}{\pi(a_1 r_1^2 + a_2 r_2^2 + a_3 r_3^2)} e_w - \frac{4(a_1 r_1^3 + a_2 r_2^3 + a_3 r_3^3)}{3v(a_1 r_1^2 + a_2 r_2^2 + a_3 r_3^2)} e_p \right)$$

$$\bigg/ \left(\frac{2}{\pi(a_1 r_1^2 + a_2 r_2^2 + a_3 r_3^2)} + \frac{4(a_1 r_1^3 + a_2 r_2^3 + a_3 r_3^3)}{3v(a_1 r_1^2 + a_2 r_2^2 + a_3 r_3^2)} \right)$$

$$\tag{1.32}$$

where

$$F_i(r_i) = \frac{\pi a_i(v/f_i) \left[v/f_i - (r_i + v/f_i) e^{-r_i f_i/v} \right]}{a_i \pi r_i^2 / 2} \tag{1.33}$$

Stationary prey   If the prey are stationary, such as nearly sessile insects or plant parts like fruits and seeds, then the capture fractions, $F_i(r_i)$, are identically one. The equations for the optimal cutoff radii then become analytically tractable. On differentiating E/T with respect to each r_i, and after considerable rearrangement, we obtain

$$r_2 = \left(\frac{e_2}{e_1} \right) r_1 \tag{1.34}$$

$$r_3 = \left(\frac{e_3}{e_1} \right) r_1 \tag{1.35}$$

where r_1 is found from a familiar equation (cf. Equation 1.14)

$$\pi Q r_1^3 + 6(e_p - e_w)r_1 - 3e_1 v = 0 \tag{1.36}$$

although here Q depends on the abundance of all three prey types

$$Q \equiv a_1 e_1 + a_2 e_2^3/e_1^2 + a_3 e_3^3/e_1^2 \tag{1.37}$$

The solution for r_1 is therefore identical to Equation 1.16 using the appropriate Q.

These formulae show that for stationary prey the fraction of diet in each prey category is independent of the lizard's SVL. The amount of material captured from prey category i scales as $a_i r_i^2$, because the area being swept up by the lizard varies as the foraging cutoff radius squared, which, in turn, is multiplied by the abundance per unit area to obtain total consumption. Therefore, the optimal diet is distributed across the prey classes in the ratios of

$$a_1 : a_2 (e_2/e_1)^2 : a_3 (e_3/e_1)^2 \tag{1.38}$$

These ratios are independent of lizard properties. They depend only on insect properties such as abundance and length. Thus, for sessile prey, lizards should have the same dietary proportions independent of body size. And in particular, a lizard's average prey size should not vary with body size.

Basis for prey size–body size relation   Yet we know that the average prey size *does* increase with lizard body size. Therefore, prey escape assumes even more theoretical importance than previously, where it was introduced solely as an aid to improve the theory's accuracy. Now we see that prey escape underlies the existence of a positive correlation between insect prey size and lizard body size. Unless large lizards have some greater ability to capture large insects than small lizards have, such as their being able to sprint faster, they should not prefer large prey any more than small lizards do. Thus, in the absence of prey escape, large lizards would consume more total prey than small lizards, but the composition of the diet would be the same for all lizard body sizes.

Biomechanical constraints could play a similar role to sprint speed. A lizard probably cannot consume items larger than about one-quarter of its SVL and obviously will ignore such items. Hence, more large prey will be taken by large lizards than by small lizards for this reason alone. Nonetheless, the data from St. Eustatius suggest that the maximum prey size found in a lizard's stomach is usually much smaller than it could physically consume, implying that this biomechanical constraint is of secondary importance. Also, we have assumed that a lizard can gobble its prey in one swallow. Yet anoles are often observed spending five minutes or more chewing and swallowing giant prey including butterflies, cockroachs, grasshoppers, walking sticks, and cicadas. Handling time might be assumed to increase exponentially with prey length, using an exponential rate coefficient that depends on the lizard's SVL. Large lizards might be assumed to handle large prey more quickly than small lizards. The key point is that a large lizard must have some *advantage* in getting large prey relative to a small lizard to explain why large lizards prefer large prey more than small lizards do. Otherwise both large and small lizards should both prefer large prey over small prey to an equal extent.

Mobile prey   To understand the impact of prey flight with multiple prey, we need to find the optimum cutoff distances, r_1, r_2, and r_3 numerically. The Newton-Raphson algorithm used previously can be extended to find roots to a system of simultaneous nonlinear equations. Vector and matrix multiplication are used to simplify the notation. Suppose $f(x)$ is a system of k equations, where x is a column vector of dimension k. Let x_0 be the initial guess at where the root is, and x_n be the approximation of the root after n iterations. Similarly, let f_n be the system of equations evaluated at x_n; this too is a column vector of dimension k. Lastly, let J_n

be the $k \times k$ matrix of first derivatives, $\partial f_i / \partial x_j$ evaluated at x_n. The multivariate Newton-Raphson algorithm then is

$$x_{n+1} = x_n - J_n^{-1} f_n \tag{1.39}$$

where J_n^{-1} means the inverse of J_n. Notice the similarity to Equation 1.25 that was used for one prey type. In this example f_n consists of $\partial(E/T)/\partial r_1$, $\partial(E/T)/\partial r_2$, $\partial(E/T)/\partial r_3$, evaluated at the nth guess for the optimal foraging radii, r_1, r_2, and r_3. The elements of J_n are quite messy, but represent a routine calculation nonetheless, and are not detailed here. The formulae for both f_n and J_n are contained in the program forage.scm that is on the disk associated with the book. Also, the program includes the functions to do elementary matrix and vector operations with 3×3 matrices. An initial guess may be the solution for stationary prey, or the distance a lizard sprints before the insect is expected to fly away, whichever is smaller.

Predictions from this optimal foraging theory for three prey types, including the possibility of prey escape, are presented in Figure 1.16. The example uses prey lengths of 1, 4, and 10 mm corresponding to the prey categories of Tables 1.15 and 1.16, based on St. Eustatius data. The total prey abundance is the familiar value of 480 insects per m^2 per 12 h. The distribution of these prey into the length categories is 96%, 3.5%, and 0.5%, based on Table 1.16. The insect flightiness coefficients for these three classes are assumed to be 1, in accordance with the assumption that a lizard has about 1 second in which to catch a prey item.

1.3.8 Optimal cutoff radii and yield

Figure 1.16 (top) shows the radii predicted for the three prey sizes. The lowest curve is for the ants. Depending on its SVL, a lizard should not travel much further than about 0.75 m to catch an ant. A lizard should go up to 3 m for a fly, and to about 7 m for a cricket.

Figure 1.16 (bottom) shows the yield to a lizard that follows this optimal strategy, depending on its SVL. The best SVL is 125 mm, which yields the energy equivalent of about 1 egg per 12 h of continuous optimal foraging. A 40-mm lizard, such as *A. schwartzi*, is predicted to have a foraging yield of about 1/3 egg per 12 h of continuous optimal foraging, and a 55-mm lizard, such as *A. bimaculatus*, is predicted to have a yield of about 1/2 egg per 12 h of continuous optimal foraging.

1.3.9 Optimal prey size–body size relation

Figure 1.17 shows the average prey length in the diet of an optimally foraging lizard as a function of its SVL. The average prey length in a 40-mm lizard is about 1.5 mm, and for a 55-mm lizard about 1.9 mm. Comparison with Table 1.15 shows these predictions are slightly low both for *A. schwartzi*, which has an average prey length in its diet of 2.3 mm, and for *A. bimaculatus*, whose average prey length is 3.0 mm.

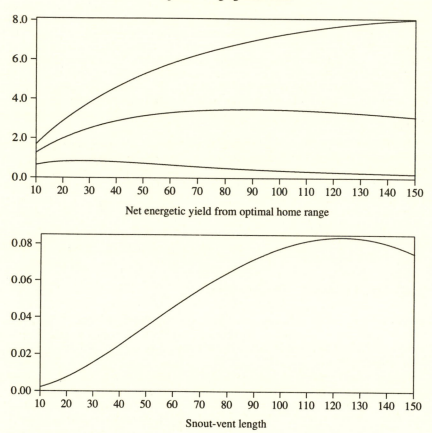

Fig. 1.16 Top: Optimal cutoff radii in m for simultaneous foraging on insects of three lengths. Curves for 10, 4, and 1 mm insects from top to bottom. Bottom: Net energetic yield, E/T in joules/s for optimal forager. Total prey abundance is 480 insects appearing per m^2 per 12 h; distribution of insects is 0.005, 0.035, and 0.960 for 10, 4, and 1 mm lengths, and insect flightiness coefficients are 1, 1, and 1.

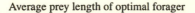

Average prey length of optimal forager

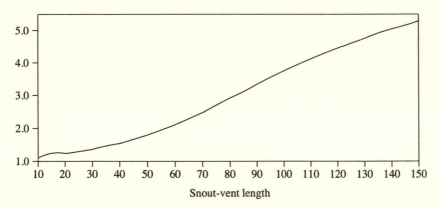

Snout-vent length

Fig. 1.17 Average prey length in millimeters for diet of optimal forager. Total prey abundance is 480 insects appearing per m^2 per 12 h; distribution of insects is 0.005, 0.035, and 0.960 for 10-, 4-, and 1-mm lengths, and insect flightiness coefficients are 1, 1, and 1.

Optimal diet

Turning now to the total number of prey caught per 12 h, consider how the diet can be predicted from this three-prey theory. The average waiting time per item is

$$t_w = \frac{2}{\pi(a_1 r_1^2 + a_2 r_2^2 + a_3 r_3^2)} \tag{1.40}$$

and the average pursuit time per item is

$$t_p = \frac{4(a_1 r_1^3 + a_2 r_2^3 + a_3 r_3^3)}{3v(a_1 r_1^2 + a_2 r_2^2 + a_3 r_3^2)} \tag{1.41}$$

As before, the total number of items pursued during period, T, is

$$N_T = \frac{T}{t_w + t_p} \tag{1.42}$$

The fraction of pursuits that are of type-i, P_i, is

$$P_i = \frac{\pi a_i r_i^2/2}{\pi a_1 r_1^2/2 + \pi a_2 r_2^2/2 + \pi a_3 r_3^2/2} \tag{1.43}$$

And the fraction of pursuits of type-i that are successful, F_i, is, as before,

$$F_i(r_i) = \frac{\pi a_i(v/f_i)\left[v/f_i - (r_i + v/f_i)\,e^{-r_i f_i/v}\right]}{a_i \pi r_i^2/2} \tag{1.44}$$

Therefore, the diet, i.e., number of prey type-i caught during the period, $\mathcal{T}$, is

$$D_i = F_i P_i N_{\mathcal{T}}, \tag{1.45}$$

the proportion of type-i in the diet is

$$d_i = D_i/(D_1 + D_2 + D_3), \tag{1.46}$$

and the average prey length, as illustrated in Figure 1.17, is

$$\bar{l} = d_1 l_1 + d_2 l_2 + d_3 l_3, \tag{1.47}$$

where l_i is the length of an insect of type-i.

These formulae work out to the prediction that a 40-mm lizard should catch 259, 36, and 6 insects of length 1, 4, and 10 mm, respectively, over 12 h of continuous optimal foraging, corresponding to an average prey length of 1.5 mm, and that a 55-mm lizard should catch 235, 62, and 12 insects of these sizes, corresponding to an average prey length of 1.9 mm, as mentioned before. These predictions assume an insect distribution of 96.0%, 3.5%, and 0.5% for the lengths of 1, 4, and 10 mm, respectively, a total insect abundance of 480 insects per m^2 per 12 h, and insect flightiness coefficients of 1 for the three insect types.

Thus, the catch per 12 h of continuous optimal foraging is predicted to be about 300 insects in the three-prey model. This prediction is much better than the prediction of 850 per 12 h from the preceding one-prey model. The new prediction may still be high though. If the lizard is foraging for about 3.5 h per day, as assumed in the next section, then the daily catch is about 75 insects, whereas a daily catch of 50 insects seems more realistic. This discrepancy may be genuine, or may reflect an underestimate in the field data of the number of small insects actually consumed or an inaccurate estimate of the relative proportions of small, medium, and large insects actually present in the environment.

In sum, then, the three-prey model appears to predict an average prey size that is somewhat low and a total daily catch that is somewhat high. These errors may be related. If the lizard were to catch fewer small insects than predicted, then both its average prey size would be larger and its total catch would be smaller. Still, I'm basically satisfied with the agreement between theory and data at this stage. While certainly imperfect, the foraging model now seems to predict much of the behavior of an anole from its physiology together with some environmental facts. To take the model further requires more accurate information on model parameters than presently available, together with further tests of the model's assumptions. Still, we have come quite far and can do much with what we now have.

1.3.10 Learning with three mobile prey types

As a result of the preceding theory, our sentient lizard now faces the need to learn much more than before. It now must learn three cutoff distances for three prey types, taking into account the possibility that prey can escape. Is our lizard up to this task?

The lizard is assumed to know how to distinguish the different types of prey, including the energy content of each and the odds of making a successful capture. As before, at each appearance of an insect the rule of thumb is to do whichever

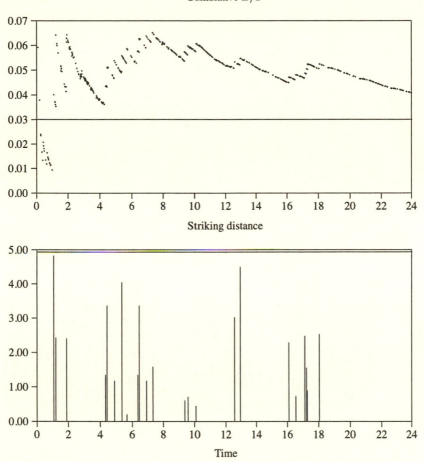

Fig. 1.18 Top: Net energy gained per elapsed time in joules/s during 24 h of simultaneous foraging on three prey sizes, as noted after each prey capture. Prey sizes are 10, 4, and 1 mm, distributed as 0.005, 0.035, and 0.960, respectively, and with flightiness coefficients of 1, 1, and 1. Bottom: Distance in meters of successful strikes for 10-mm prey.

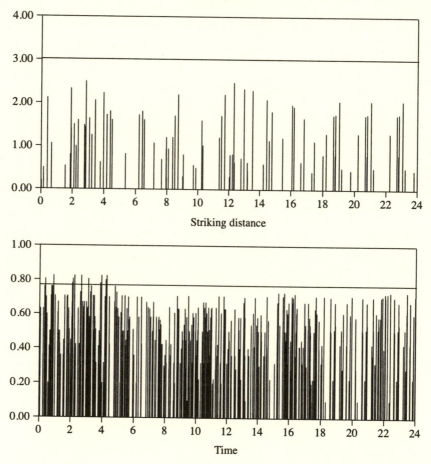

Fig. 1.19 Top: Distance in meters of successful strikes for 4-mm prey. Bottom: Distance in meters of successful strikes for 1-mm prey. Abundance is 480 total insects per m² per h. Lizard SVL is 45 mm, and lizard's memory extends to beginning of foraging period. From forage.scm using xmax = 15, ymax = 15, look-xy-step = 0.1, look-t-step = 0.1, and nbig = 1000. Run with setting (randomize) to system clock.

action leads to a higher projected E/T. That is, when an insect appears, the projected E/T, if the item is pursued, is

$$(E/T)_p = \frac{E + p_j e_j - t_w e_w - t_p e_p}{T + t_w + t_p} \tag{1.48}$$

and the projected E/T, if the item is ignored, is

$$(E/T)_i = \frac{E - t_w e_w}{T + t_w} \tag{1.49}$$

where j is a subscript for the type of prey, $j = 1, 2, 3$, that has just appeared. The rule is to do whichever action leads to

$$\max[(E/T)_p, (E/T)_i] \tag{1.50}$$

Figures 1.18 and 1.19 illustrate that a sentient lizard can learn how far to chase three prey types. Figure 1.18 (top) illustrates the cumulative E/T for 24 h of foraging. Figure 1.18 (bottom) and Figure 1.19 (top and bottom) illustrate the distances at which prey of the three types were successfully pursued during this period. The predicted optimal cutoff distance is shown as a horizontal line in these graphs. A period of 24 h is used for this simulation, compared with the much shorter times in the previous simulations involving one prey type because of the long time needed for the rare prey type (the 10-mm size) to be experienced. In fact, the lizard consistently overvalues the most common type during the first 4 hours. It caught only three 10-mm insects during this period. Then it caught nine large prey during the next 4 hours, resulting in a high overall E/T. The trace of the cumulative E/T shows the dominant influence of the rare captures of the 10-mm insects. The E/T curve appears to be made up of descending segments. A new segment is started whenever a large prey is caught. In fact, over the 24 h of the simulation, the trace of E/T did not approach its asymptotic value as closely as the trace for one prey does in a few minutes. The simulation was not carried beyond 24 h because it does not seem realistic to imagine that a lizard's memory of prey capture extends in detail much beyond two full foraging days, although a longer simulation could of course be carried out in the future if needed.

The computer program on the diskette, forage.scm, was used for the simulations. Unlike previous programs, however, it is not short and simple but extends over 25 single-spaced pages.

Size escape — being small

While our emphasis has been on lizards, the theory offers an implication for insects: How to escape predation. The best way is simply to be small. A lizard ignores small- and even medium-sized prey at the margin of its home range. An ant, for example, is in jeopardy only if it is quite near a lizard, otherwise it will be ignored even though it is within a lizard's home range. The idea of a "size escape" is well known in ecology and usually refers to prey that are too big for a predator to catch or to consume. Here, the size escape is being too small to be wanted.

The overall effect of this size escape is that lots of insects survive the gauntlet of lizard predation. These underused prey are then available to other species of predators with different cost/benefit accounts, such as spiders, or even a second lizard species; or they may get off scot-free.

Table 1.17 Abundance of *Anolis* Lizards, St. Martin

	Boundary		Pic du Paradis	
Date	*A. ging.*	*A. pogus*	*A. ging.*	*A. pogus*
Jul 1977	50.1 (2.2)	0	8.4 (2.6)	10.5 (0.7)
Jul 1978	65.0 (3.0)	0	14.2 (5.5)	24.9 (3.3)
Jul 1979	64.8 (2.8)	0	7.6 (6.2)	21.2 (2.0)
Mar 1980	129.8 (5.5)	0	14.6 (3.1)	43.9 (5.2)
Nov 1980	95.4 (3.0)	4.4 (1.8)	39.8 (4.7)	56.8 (2.7)
Mar 1981	72.1 (3.3)	2.0 (1.0)	13.2 (2.9)	39.4 (2.5)
Oct 1981	122.3 (6.5)	1.3 (0.7)	15.7 (4.8)	54.3 (6.1)
Jan 1983	111.7 (3.5)	2.0 (0)	18.6 (2.7)	44.0 (2.2)
Jun 1983	54.4 (1.9)	0	14.4 (1.8)	30.6 (3.5)
Mar 1984	114.7 (3.5)	0.5 (0)	14.6 (1.5)	41.8 (1.2)
Aug 1984	49.3 (2.3)	0	6.4 (1.4)	33.0 (1.3)
Mar 1985	127.5 (5.5)	0	12.7 (1.5)	43.9 (1.9)
Jun 1985	72.8 (7.1)	0		
Apr 1986	86.0 (6.6)	0	16.4 (2.7)	48.1 (3.5)
Sep 1987	110.0 (9.7)	0.5 (0)		53.1 (4.9)

1.4   Seasonality and growth, St. Martin

The way a lizard grows throughout its life somehow reflects the accumulation of its day-to-day foraging. Perhaps, then, the daily foraging theory just developed can be extended to predict how a lizard grows throughout its entire life. To move beyond the minute-to-minute time scale of foraging theory to the monthly and even yearly scale of life history phenomena, data from the Lesser Antilles, primarily St. Martin, are reviewed. These data show how the abundance of anoles varies throughout the year, how anoles grow, and when they start reproducing.   These data are also compared with similar studies from Grenada, Panama, and elsewhere.

St. Martin in the northeastern Caribbean is the central island of the Anguilla Bank, with Anguilla to its north and St. Barths to its south. It has two *Anolis* species, *A. gingivinus* and *A. pogus*. *Anolis gingivinus* is found throughout the Anguilla Bank, whereas *Anolis pogus* is restricted to the hills in the center of St. Martin. Thus, most locations on St. Martin (and all of Anguilla and St. Barths) have only *A. gingivinus*, whereas the central hills of St. Martin have both species.

Abundance

Table 1.17 presents the abundance of anoles on St. Martin from two locations based on up to 15 census dates spanning 10 years from July 1977 to September 1987. The Boundary site is near sea level on the leeward side (western side) of the island and is located near the political boundary between the French and Dutch jurisdictions. It is relatively xeric, with abundant cacti interspersed among trees that can attain a height of 6 m. *A. gingivinus* is usually the only anole at the site, although a few *A. pogus* are seen in the winter. The Pic du Paradis site is at the highest point on the

Table 1.18   Seasonal Abundance of *Anolis* Lizards, St. Martin

Species	Center	Amplitude	Phase
	Boundary		
A. gingivinus	94.8	32.6	-0.36
A. pogus	1.1	1.6	-1.34
	Pic du Paradis		
Species	Center	Amplitude	Phase
A. gingivinus	16.9	8.8	-1.31
A. pogus	41.8	16.8	-0.92

$$n = a + b\cos(2\pi(t - 1 - c)/12)$$

where:

n is Number per 100 m^2, t is Month,

a is Center, b is Amplitude, and c is Phase

island, a hill of 424-m elevation. It is the most mesic spot on the island, wooded with trees that attain a maximum elevation of 8 m. Here both *A. gingivinus* and *A. pogus* coexist year round, and *A. pogus* is the more abundant.

Figure 1.20 illustrates abundance plotted against the month during which the census was taken. The figure reveals the clear seasonality characteristic of the northern Lesser Antilles, with the anole abundance nearly doubling between the summer and winter. The seasonality is summarized in Table 1.18 in the form of a fit of a cosine function to the data. For example, at Boundary the abundance of *A. gingivinus* at month t is $94.8 + 32.6\cos[2\pi(t - 1 + 0.36)]$. The phase of -0.36 indicates that the abundance peaks 0.36 of a month before January (which is month 1).

Figure 1.20 also illustrates a remarkable year-to-year constancy. The abundance during the summer is almost identical among years, as is the abundance among winters, although there is slightly more error in the winter estimates. The maximum life span of these anoles is about 2 to 3 years, and the typical life span is probably about 1 year, as discussed later, so that these 10 years of observation span several generations and complete turnovers of the populations. These data indicate that the anoles here have a steady-state population size.

This observation is important especially in contrast with the dynamics of continental anoles [326] and many other animal populations. In a 10-year census record of *Anolis limifrons* from Panama, on the Smithsonian's site at Barro Colorado Island in the Panama Canal Zone, the abundance varied by a factor of over 5 between 1971 to 1981, from 1 per 58 m^2 to 1 per 10 m^2 [10]. Even at its most abundant, this anole is about one-tenth the abundance of similar anoles on eastern Caribbean Islands. Also, *Anolis limifrons*' survival was 74% per 28 days regardless of sex or time of year, and did not vary over the 10-year period. The general conclusion was that the population size of *Anolis limifrons* in Panama is the sum of variable egg production and a relatively constant loss due to predation [327].

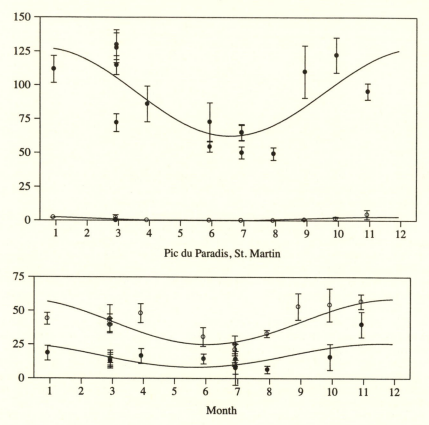

Fig. 1.20 Abundance of *Anolis gingivinus* (●) and *Anolis pogus* (○) at the Boundary site (top) and the Pic du Paradis site (bottom) on St. Martin expressed in number per 100 m² as a function of month in year. Data from dates spanning July 1977 through September 1987 (number of census dates is 15 for Boundary, 14 for Pic du Paradis *A. pogus*, and 13 for Pic du Paradis *A. gingivinus*). Error bars indicate ± two standard errors, illustrating an approximate 95% confidence interval. Curves are best fit to cosine function using the simplex algorithm; coefficients in Table 1.18. (*A. pogus* at Boundary rounded to 0 when negative.)

Table 1.19 Seasonality in St. Kitts.

	Center	Amplitude	Phase
Rainfall	89.0	58.7	-2.86
Temperature	26.7	1.6	-5.30

Seasonality

Figure 1.21 illustrates the seasonality of rainfall and temperature as recorded at the airport of St. Kitts, a close neighbor of St. Martin, during the six years between 1972 and 1977. Table 1.19 shows the coefficients in a fit to a cosine function. The rainy season can be considered as August through November.

Taken together, Figures 1.20 and 1.21 show that the rise in abundance of lizards during the winter months follows the rainy season by about 2.5 months, because the lizard abundance peaks 0.36 months before January for *A. gingivinus* at Boundary, and the rainfall peaks 2.86 months before January, implying a 2.5-month difference between lizard and rainfall seasonality. The winter's high abundance is greeted by a crunch during March and April when the dry season is in full force. The drop in abundance during the late spring could, in principle, reflect either an increase in mortality or an absence of reproduction (or both) at that time. If the situation is analogous to Panama though, the mortality is fairly constant throughout the year, and the seasonal cycles would primarily reflect seasonal cycles in reproduction.

Seasonality in the southern Lesser Antilles follows a similar pattern, although the rainy season appears to start two months earlier, in June, and to last one month longer, though December [353]. Weekly insect samples collected in Grenada during 1977 show the seasonality in total insect abundance lagging the rainy season by about three weeks [363]. The average rainy season abundance was 2.3 times greater than the dry season abundance. Also, there was no seasonal difference in insect sizes, and many taxa showed parallel responses to rainfall. Mating observations of *A. aeneus* were recorded during during October 1972 to August 1973 at a site in Grenada as 1 in April, 3 in May, 3 in June, and 1 in August [353]. The percentage of gravid females went from a low of about 10% in February to 50% in April to over 90% in June. The males showed about the same pattern, but not as markedly. About 50% of the males had sperm at the low in February and nearly 100% in May. These data suggest that the onset of lizard reproduction in Grenada matches the onset of the rainy season.

In addition to any broad differences that may exist in the timing of seasonality between the southern and northern Lesser Antilles, as indicated by the differences in seasonality between Grenada and St. Kitts, there are also differences between sites within an island. The west of St. Croix is relatively wet, with about 125 cm annual rainfall, while the east has only 75 cm annual rainfall. The rainy season is considered from September through December and the dry season from February through July [36]. In February 1979 only 8% of the female *Anolis acutus* from samples collected in the east of the island were reproductive, while 77% of the females in the west were reproductive [279]. By May though, nearly 100% of the anoles from both sides of the island were reproductive. The cumulative annual reproduction was thus much higher on the western side compared with the eastern side of the island, and the western side of St. Croix was reproductive year round.

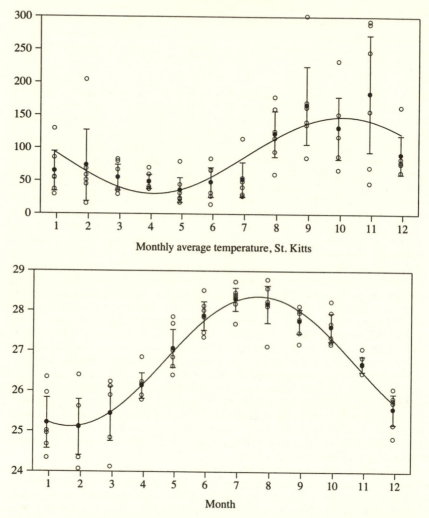

Fig. 1.21 Top: Rainfall in millimeters per month. Bottom: Average temperature each month in °C. Data recorded at airport on St. Kitts from January 1972 through December 1977. Solid dots indicate six-year average, and error bars indicate ± two standard errors.

Seasonality on the continent, in Panama, is also similar to the Lesser Antilles in many respects. The insect abundance of most insect groups track the wet season in parallel, with insect size changing very little during the year [183]; and reproductive activities of anoles are concentrated in the early rainy season [329, 328]. Thus, the main differences between Panamanian populations of anoles and those of the Lesser Antilles appear to be that (1) Panamanian anoles are less abundant than Lesser Antillean anoles by an factor of 10 or more; (2) Panamanian anoles show interannual variability of a factor of 5 or more between "good" and "bad" years; (3) the mortality rate in Panama is very high, though apparently not variable seasonally or interannually; and (4) as mentioned again below, in Panama insect prey size is larger, individual growth rates are higher, and time to reproduction is earlier than in the Lesser Antilles [5].

Growth

During experiments to investigate competition between the species of St. Martin, data were obtained on the growth rates of individuals. The experimental enclosures were constructed at a site approximately midway between Boundary and Pic du Paradis, at approximately 200 m elevation. Outside the enclosures at this site, both species are approximately equally abundant. Anoles in experimental enclosures were individually marked and recaptured in consecutive months. Figure 1.22 shows the daily ΔSVL as a function of SVL for *A. gingivinus* from an enclosure containing 60 individuals in 144 m² area. Of this area, 64 m² in the center of the enclosure is naturally wooded, while the remaining area is a strip adjacent to the enclosure's fence that was cleared of vegetation. The number of anoles in the enclosure, 60, is a more or less natural degree of crowding for the wooded center of the quadrat, although relatively uncrowded for the total area of the enclosure, thus leading to conditions in which nearly maximal growth is expected.

The maximum growth rate observed was about 0.20 mm/d for juveniles, and the the daily growth increment drops monotonically as the lizards get bigger. Females stop growth altogether by 50 mm, and males by about 63 mm. The solid curve in the figure is almost, but not exactly, a straight line. This curve was obtained from the foraging theory assuming 3.5 h of foraging on the average, every day of a lizard's life, as derived in the next section.

A straight-line relationship between the daily ΔSVL and SVL is called the Bertalanffy model of growth [371]. The Bertalanffy model is a phenomenological description of growth that has been interpreted somehow as showing that growth is the difference between synthetic processes that scale as body mass to the 3/4 power and degradative processes that scale directly with body mass. Another phenomenological description of growth is the logistic equation, for which the relation of daily ΔSVL to SVL is a parabola. Both phenomenological models have been used to describe growth in reptiles [7], and the logistic equation has tended to be favored for small lizards [5, 316, 99].

The St. Martin data, however, point to a Bertalanffy model, rather than a logistic model, because the data are more readily viewed as lying on a descending straight line than on the hump-shaped curve of a parabola. Possibly the absence of specimens from 15 mm to 30 mm simply does not reveal the left side of a hump-shaped curve, and that the data happen to lie on the right side of a descending parabola after all. Still, there is presently no evidence for logistic growth in these data that seem instead to support a Bertalanffy model.

Daily growth

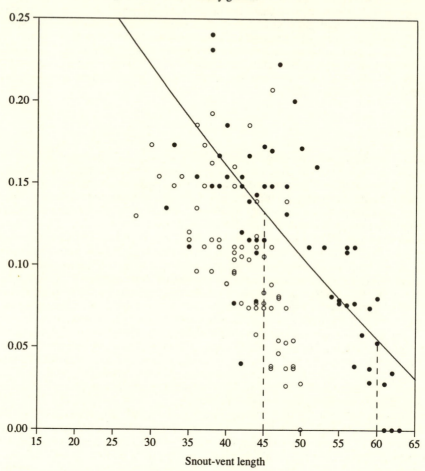

Fig. 1.22 Growth of *Anolis gingivinus* in mm/d for males (●) and females (○) in St. Martin from December 1980 to April 1981 as a function of a lizard's snout-vent length in mm. Solid descending curve is growth predicted from optimal foraging theory assuming 3.5 h foraging time per d and using the optimal foraging parameters of Figure 1.16. Vertical dashed lines are conjectured switches from growth to reproduction, for females at 45 mm and males at 60 mm.

Because the solid curve derived in the next section from foraging theory is nearly a straight line, this theory can be considered as offering a derivation of a Bertalanffy-like model from the first principles of optimal foraging theory.

Data are available for the growth rates of hatchlings and very small juveniles, those with SVL's between 19 mm and 30 mm, for *Anolis aeneus* of Grenada [346, 356]. The data vary from 0 to almost 0.30 mm per day depending on the rainfall. This study is important in showing that much of the variation in daily growth rates is correlated with rainfall. Rainfall is correlated with insect abundance, and water is directly needed by the lizards; both factors were shown to contribute to explaining variability in growth rates. Also, this study offers evidence against a logistic model of growth, with the finding that daily growth rates were unrelated to SVL for anoles in this size range. If the logistic model were true, a strong positive relation (left side of the hump-shaped curve) should be observed between daily growth and SVL. Finally, these studies showed not only a correlation of daily growth rates with rainfall but also that growth rates are increased when natural sites are experimentally watered.

Also on Grenada, the chronology of egg production has been determined [345]. Enlarged follicles are first detectable at 2-mm size by gently palpating the animal. From this point the follicle is ovulated in 10 days when the follicle is 6 mm. It is laid after 5 or 6 more days (15 to 16 days total) when the egg is 10 mm in diameter. However the exact timing of egg laying depends on rainfall, and a female can retain an egg while awaiting rain. Egg-laying behavior is induced both by natural rain squalls and by experimental watering of a study site. Females may dig holes, and not lay eggs in them, presumably in anticipation of egg laying. Stamps describes two distinct behavior patterns when eggs are laid: "One, digging, involved scraping of the ground with the forepaws. Generally, females dug with one paw for several strokes (4 to 11) then shifted to the other paw. A second pattern, poking, involved insertion of the tip of the snout into a prospective hole site or into a partially dug hole. Poking usually occurred periodically during hole digging." I have also witnessed these behaviors in *Anolis gingivinus* on the cay, Anguillita, at the eastern tip of Anguilla. Stamps reported that a female laid an egg on each of two consecutive days, implying that an egg may be retained while awaiting rain, and also that two eggs were never laid in the same hole. Eggs laid in soil that is too dry die of dessication [329, 242].

The growth rates of anoles in Central America appear to be higher on average than anoles on Caribbean islands and do not increase with addition of water or food [5]. The *maximal* growth rates in Central America are no higher than maximal rates in the Caribbean; instead, the finding is that under usual field conditions the growth rates are higher in Panama. The interpretation is that Caribbean anoles are more resource-limited than Central American anoles. Food limitation, if any, in Central American anoles may come during the dry season, when growth rates are slower than during the rainy season [11]. Insect prey sizes are also higher in Central America [5]. As a result of their high growth rates, Central American anoles grow to maturity in two to four months, quicker than the five to nine months of Lesser Antillean anoles [5]. This advantage is counteracted by the high mortality, leading overall to a high population turnover rate.

Life history

The life history of a female anole has three episodes: the hatchling to juvenile phase during which it is solely growing, a transition phase during which it is both growing and producing eggs, and a mature phase when growth has completely ceased and all the energy from foraging is allocated to reproductive activities.

In Grenada the smallest female size exhibiting reproductive activity is 39 mm [343], and in St. Croix the smallest reproductive size is also 39 mm [296]. The largest female sizes are 51 mm in *A. aeneus* of Grenada and 49 mm in *A. acutus* of St. Croix, respectively.

Figure 1.22 can be interpreted as showing that female *A. gingivinus* of St. Martin grow quickly from hatchling to about 40 mm, grow more slowly between 40 mm and 50 mm while reproduction begins, and stop growth altogether at 50 mm. While there are scattered reports that females grow more slowly than males from the time they hatch [7], Figure 1.22 suggests that both sexes start out at the same growth rate.

Thus, a provisional interpretation is that both males and females start out with the same growth curve, that females undergo a transition to maturity between the sizes of 40 mm and 50 mm and that males undergo their transition between 55 mm and 65 mm. The center of the transition for females is about 45 mm, and for males about 60 mm. These sizes are indicated with vertical dashed lines in Figure 1.22. Therefore a somewhat idealized view of the life history is that both males and females start growing along the same solid Bertalanffy-like curve, and then females switch off it at 45 mm while males continue along the curve until 60 mm when they too switch off. The objective of the life history theory to follow is to predict when this switch occurs.

In the future, for better accuracy, the transition to maturity can be explicitly considered as a graded transition instead of discrete switch, as with the gradual transition from vegetative growth to flowering in annual plants [165] or as sequential switches between vegetative growth and reproduction in successive seasons of a perennial plant [172, 171]

Two other issues needing future consideration are the roles of fat storage and ontogenetic habitat shifts. A simplification of anole growth and development is to assume that the yield from foraging is allocated either to growth or to reproduction, whereas in reality, a third possibility exists—allocation to storage as fat bodies. Anoles can accumulate fat deposits during the dry season and draw them down during reproduction [186, 346, 279]. Life history allocation theory for plants has also considered the role of storage in structures such as roots [159, 61] and can presumably be adapted for fat storage in lizards. Another simplification has been to split out the time used for foraging from time spent watching out for predators and in social activities. Also, predation may lead to shifts in habitat with age [348, 349, 350], similar to ontogenetic shifts in habitat by fish [374, 373].

1.5 Growth of an optimal forager

Suppose a lizard wakes up each morning and forages optimally for that day in accordance with the body size it has at the start of the day. Then at night, when the lizard is asleep, it grows a little bit by absorbing into its biomass the net yield it has just achieved from its foraging. The next morning when the lizard reawakens, it is a little longer than it was the day before, and forages optimally on this day according to the new body size it now has.[5] If this process is repeated day after day, gradually the lizard grows from its starting size as a hatchling into the size

---

[5]Incidentally, the sleeping behavior of anoles on Puerto Rico is studied in [62]

range for adults. The question is: Does this model predict a growth curve that resembles the actual growth curve for anoles?

Each day we first calculate the lizard's mass from its SVL according to the formula from Table 1.9,

$$M = (3.25 \times 10^{-5})L^{2.98} \tag{1.51}$$

where L denotes the SVL in mm and M is the mass in g. After 24 h we assume the lizard has acquired some net energy, say ΔE, in joules. We convert this ΔE into a change in mass, ΔM, in grams wet weight, with the formula

$$\Delta M = \Delta E / 6300 \tag{1.52}$$

where the conversion constant is 6300 joules per gram [167, 229]. The M is then added to its mass to obtain a new mass, which is then converted back to body length by

$$L = \left[M / (3.25 \times 10^{-5}) \right]^{1/2.98} \tag{1.53}$$

Thus, each day's net yield of foraging energy leads to a little bit of growth.

The main subtlety is in determining how much foraging is carried out per 24 hours for each and every day in a lizard's life. On a good day anoles start activity about 1 to 2 h after sunrise and continue to sunset. On especially hot days they may have a midday period of inactivity, and on cold and rainy days they may not be active at all. The likelihood of these situations varies with site and season. Moreover, even when a lizard is active for a large part of the day, it is not known what fraction of this time is spent foraging in the sense of observing and responding to the appearance of prey. When a predatory bird such as a Pearly-Eyed Thrasher is nearby, an anole keeps a wary eye on it and ceases foraging altogether. Also, territorial interactions and courtship consume a significant fraction of an anole's activity. So we will simply guess that an anole in typical xeric sea-level Lesser Antillean habitat spends foraging about 3.5 h per day, every day on the average. This number, as we will see, leads to plausible predictions: It can be thought of as another parameter estimated from the data with which the model is being compared, and reducing the model's degrees of freedom by one.

For a specified foraging time, t_f per 24 h, the yield per 24 h, is

$$\Delta E = 86400 \left(\frac{t_f}{24} E/T_o(L) - (1 - \frac{t_f}{24})e_r \right) \tag{1.54}$$

where $E/T_o(L)$ is the yield in joules per second for an optimal forager as a function of its snout-vent length (Figure 1.16, bottom), and e_r is the lizard's resting metabolic rate in joules per second (Table 1.9). The 86,400 in the equation simply converts the units of time from seconds into days.

Figure 1.23 illustrates the growth curve predicted for an optimal forager, assuming 3.5 h foraging per day in the ecological circumstances previously used for the optimal foraging predictions of Figure 1.16. The computations for this figure were obtained from the program `grow.scm` on the diskette.

The predicted growth curve simply asymptotes at a snout-vent length of about 70 mm, which is slightly larger than the maximum size of males in sea-level habitat on islands with one species of anoles. With longer activity times per day, the lizard asymptotes at a larger size, and conversely for shorter activity times. The value of

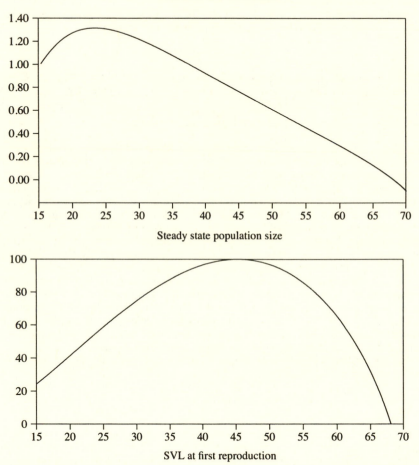

Fig. 1.23 Snout-vent length in mm as a function of age in days assuming 3.5 hrs foraging hours per day, every day.

3.5 h foraging per day was selected to agree with the daily growth data of Figure 1.22 while keeping all the same ecological parameters that were already used when developing the three-prey foraging theory predictions.

The predicted growth curve is surprisingly simple. The daily ΔSVL is very close to being a linear decreasing function of SVL, which is the Bertalanffy growth model, as mentioned earlier. Future theoretical efforts may be able to take advantage of this simplification.

1.6 Optimal life history of an optimal forager

The age at which to stop growing and to begin reproduction can be determined, in principle, from life history theory. The idea is to find the particular age for switching from growth to reproduction that maximizes an individual's contribution of descendants to future generations. This age is called the optimal switching age. If there is variation in age for switching that is inherited without too many genetical complications, then natural selection will bring about the evolution of the "optimal switching age". The problem is to determine what age of switching really is best in terms of producing the largest number of descendants. The answer to this problem may depend, and indeed appears to here, on whether the lizard population is expanding in abundance or at steady state.

1.6.1 Expanding population—r-selection

If the population is expanding, its exponential rate of increase, r, is known from population theory to be the root of the equation

$$\int_0^\infty e^{-rx} l(x) m(x) dx = 1 \qquad (1.55)$$

where $l(x)$ is the survivorship curve [$l(x)$ is the probability of living to age x or more] and $m(x)$ is the maternity function [$m(x)dx$ is the number of offspring produced as an animal ages from x to $x + dx$].[6] The population genetic theory usually presented in textbooks is based on a model of population dynamics that lacks age structure and does not allow overlapping generations, as occurs in annual plants. In fact, population genetic theory was extended to age-structured populations in 1928 [232]. This extension resulted in a theorem that, in an expanding population, natural selection brings about the evolution of those individuals whose life history possesses the highest r, where r is computed from Equation 1.55 above. Today, this situation is also called "density-independent natural selection" and also "r-selection" [199].

Here, we need to find the age for switching from growth to reproduction that maximizes r. Let the switch time be denoted as τ; this is the age when reproduction

---

[6]This equation, basic to demographic theory, is derived by writing out a model that predicts the number of individuals of each age through time, $n(x, t)$. Then substituting a trial solution of the form $n(x, t) = c(x)e^{rt}$ leads to Equation 1.55 for r. The $c(x)$ is called the stable age distribution. Thus, the population actually grows at rate r only when the stable age distribution has been attained. In practice, r from this formula approximately equals the actual population growth rate, and the stable age distribution approximately equals the actual age distribution, after a time has passed equal to the maximum life span.

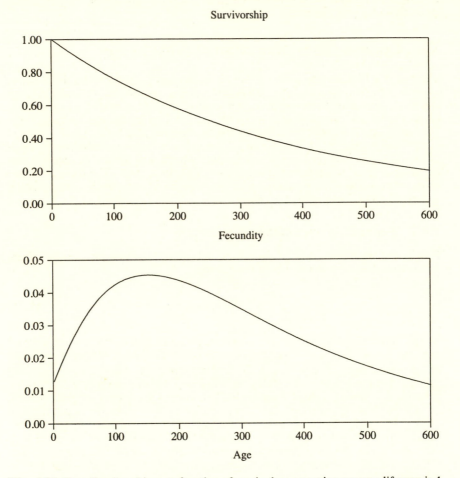

Fig. 1.24 Top: Survivorship as a function of age in days assuming average lifespan is 1 year. Bottom: Fecundity in eggs per day as a function of the age in days when reproduction is started by switching the daily foraging yield from growth to egg production. Assumes foraging is for 3.5 h per day, every day.

begins and growth stops—the age of first reproduction. We therefore need the two functions, $l(x)$ and $m(x)$ to be expressed in terms of τ. We then evaluate the integral, determine r as a function of τ, and identify the τ that leads to the highest r: This is the optimal switching age for a population growing in a density-independent situation.

The survivorship of anoles in the Bahamas as been shown to be well described by an exponentially decreasing function,

$$l(x) = e^{-\mu x} \tag{1.56}$$

with a mortality parameter μ that corresponds to an average life expectancy of 1 to 2 years [317]. We use here an average life expectancy of 1 year, as illustrated in Figure 1.24 (top). We assume this curve is independent of τ because the lizard is assumed to forage to the same extent regardless of whether it is growing or reproducing. If the predation hazard is primarily incurred during foraging, then the risk of mortality experienced is independent of whether the acquired resources are allocated to growth or reproduction.

The maternity function, $m(x)$, is a step function that equals zero from age 0 to age τ, and is a constant, $m(\tau)$, from age τ to ∞. The value of $m(\tau)$ depends on the SVL that has been attained at τ, and equals the daily foraging yield at age τ converted from joules to eggs. Figure 1.24 (bottom) illustrates $m(\tau)$. The highest potential fecundity, 0.0453 eggs per day, is attained when the lizard is 152 days old and has an SVL of 46.1 mm. This fecundity corresponds to one egg per 22 days, which is a realistic value for anoles. In summary, $m(x)$ is taken to be step function in which the step occurs at age τ and the height of the step is $m(\tau)$ as depicted in Figure 1.24 (bottom).

With $l(x)$ as an exponential function, and $m(x)$ as a step function, the integral in Equation 1.55 simplifies, leading to the following equation for r

$$m(\tau)e^{-(\mu+r)\tau} - (\mu + r) = 0 \tag{1.57}$$

This equation can be solved numerically (with the Newton-Raphson algorithm again) to yield $r(\tau)$. The best switch time is the τ that produces the highest r.

Figure 1.25 (top) shows $r(\tau)$ and illustrates that the best switch time according to this criterion is at 28 days when the lizard has grown from its hatchling size of 15 mm to 23 mm. The fecundity at this size is 0.0244 eggs per day, or 41 days per egg. This prediction is quite far off the mark: it predicts that reproduction begins much too early and at much too small an SVL. I think this prediction is falsified by the data available. The life history of anoles does not seem consistent with the predictions of r-selection theory.

1.6.2 Steady state population—K-selection

Life history theory for a steady-state population is not as secure as that for an expanding population. A steady state involves density dependence: either survivorship, fecundity, or both decline with population density. The first attempt to extend evolutionary theory to include density dependence was taken by MacArthur in 1962; he suggested that natural selection in the presence of density dependence maximizes a population's K rather than its r, where the terms r and K are drawn from the customary notation for the logistic equation.[7] MacArthur coined the

---

[7]The logistic equation is $dN/dt = rN(K - N)/K$, where N is the population size. The solution, $N(t)$ is the familiar sigmoid curve showing the population size approaching a steady state at $N = K$.

term "K-selection" for this situation [199]. During the 1970's the idea of density-dependent natural selection was developed in the explicit context of population genetic theory, and a general density-dependent counterpart of the fundamental theorem of natural selection was developed that was not tied specifically to the logistic model [280, 4, 63, 37], and more recently has been confirmed in laboratory experiments with *Drosophila* [228]. The finding from this theoretical work was that density-dependent natural selection favors, and eventually fixes, alleles that increase the steady-state population size. While the life history literature quickly adopted (and often abused) the idea of K-selection, it remained poorly known that K-selection theory (i.e., density-dependent selection theory) has no age structure in it and therefore really doesn't say very much about life histories. In fact, density-dependent selection theory extends traditional population genetics *only* in adding the density dependence and leaves traditional population genetics as is, lacking both age structure and frequency dependence.

Here though, we do need a life history theory with both density dependence and age structure. So, let us tread lightly into some new life history theory. We assume, and do not prove, that density-dependent selection tends to maximize the steady-state population size with age structure just as it does without age structure, provided the density dependence is of a simple form. We consider two cases, density-dependent survivorship and density-dependent fecundity.

Density-dependent survivorship

If survivorship is the aspect of the life history primarily influenced by density, then the survivorship curve at a steady-state abundance, $\widehat{N}$, can have the form

$$l(x, \widehat{N}) = l_o(x)e^{-a\widehat{N}} \tag{1.58}$$

where $l_o(x)$ is the density-independent component of the survivorship and a is a constant that indicates the effect on survivorship per unit of population. In this situation, the steady-state population size is found from Equation 1.55 with r equal to 0,

$$\int_0^\infty l_o(x)e^{-a\widehat{N}}m(x)dx = 1 \tag{1.59}$$

The idea is then to find the life history that maximizes $\widehat{N}$. But inspection of this equation indicates that it's the same as the equation for r except for a change in the name of what's being maximized. Therefore, the life history that maximizes r, as appropriate for an expanding population, is also the best life history for a steady-state population with density-dependent survivorship, provided a is independent of the life history parameter being selected. In particular, the best switch time by the criterion of maximizing $\widehat{N}$ remains at an age of 28 days and a size of 23 mm as depicted in Figure 1.25 (top). Therefore, we continue to be far from the mark, but matters are about to improve.

Density-dependent fecundity

If fecundity is the aspect of the life history primarily affected by population density, then the maternity function can have the form

$$m(x, \widehat{N}) = m_o(\tau) - a\widehat{N} \quad \text{if } x \geq \tau \tag{1.60}$$

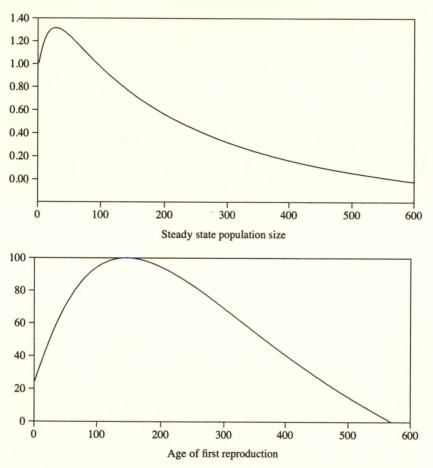

Fig. 1.25 Top: Intrinsic rate of increase, r ($\times 10^2$), as a function of the age in days at which growth stops and reproduction begins. Bottom: Steady-state population size, $\widehat{N}$, in number of lizards per 100 m[2] as a function of the age at which growth stops and reproduction begins. Foraging time is 3.5 h per d.

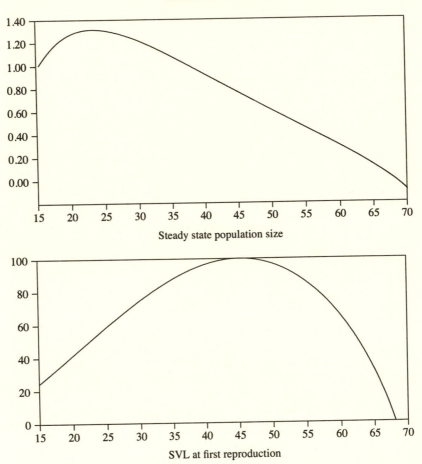

Fig. 1.26 Top: Intrinsic rate of increase, r ($\times 10^2$), as a function of SVL in mm at age of first reproduction. Bottom: Steady-state population size, $\widehat{N}$, in lizards per 100 m^2 as a function of SVL in mm at first reproduction.

$$= \quad 0 \quad \text{otherwise} \qquad (1.61)$$

where $m_o(x)$ is the density-independent component of fecundity and a is a constant indicating the effect on fecundity per unit of population. The interpretation is that the animals grow and mature in density independent conditions: the rainy season and soon thereafter when food is relatively abundant. Then the density dependence enters when they start to reproduce, resulting in egg production that decreases as a linear function of population size. The equation for the steady-state abundance now becomes

$$\int_{\tau}^{\infty} l(x)[m_o(\tau) - a\widehat{N}]dx = 1 \qquad (1.62)$$

This may be rearranged by doing the integral and solving for $\widehat{N}$ as

$$\widehat{N} = (1/a)[m_o(\tau) - \mu e^{\mu\tau}] \qquad (1.63)$$

We set the parameter a to be such that the maximum $\widehat{N}$ equals 100 lizards per 100 m^2, which we take, by convention, as a maximum steady-state population size.

Figure 1.25 (bottom) displays $\widehat{N}$ as a function of the switching time, τ. According to this criterion, the age at first reproduction should be 145 days at an SVL of 45.3 mm. The fecundity at this size is 0.0453 eggs per day, or 22 days per egg. (This is one week earlier than the age of switching that yields the highest possible fecundity.) Figure 1.26 shows the intrinsic rate of increase and the steady-state population sizes graphed against the snout-vent length at first reproduction, rather than τ, to avoid the need for continually translating between τ and SVL using Figure 1.23.

This conclusion appears consistent with the data on *A. gingivinus* from St. Martin. As mentioned earlier, the females switch from growth to reproduction at an SVL of about 45 mm, while the males switch at about 60 mm, which is smaller than the potential maximum size of about 70 mm for the environmental conditions being modeled. Also, this conclusion is consistent with the data indicating that the abundance of anoles is in steady state (cf. Figure 1.20). Thus, the life history of anoles does seem consistent with the predictions of K-selection theory provided that the density dependence primarily affects fecundity and not survivorship.

Where does density dependence act?   While there is excellent evidence that the abundance of *Anolis* populations of St. Martin are at a steady state brought about by density dependence, there is presently no evidence either way as to whether the density dependence acts on fecundity rather than survivorship.

Consider the seasonal fluctuation in abundance. It *could* be explained by density-dependent survivorship. If hatchlings are continually produced in excess of available space for territories, then once the habitat fills up, predation could remove the individuals that did not find territories: This predation would be density-dependent because its intensity would increase as space for territories is exhausted.

Alternatively, the seasonal fluctuation *could* be explained by density-dependent fecundity. There could be more or less continual predation and mortality through-out the year and a diminishing of reproduction as food becomes scarce when the summer's dry season approaches. This scenario envisions a gradual dieback of the lizard population that is restored by reproduction when the rainy season begins anew.

A comparison with Puerto Rico may shed light on where density dependence operates. *A. stratulus* catchs 15 to 18 insects per day during the dry season and 20 to 25 insects per day during the wet season and nearly twice as many moves are required to catch a given number of prey in the dry season relative to the rainy season [266, 190]. The lower total catch and higher requirement of energy per unit catch certainly lowers the daily foraging yield, and therefore the fecundity, during the dry season. Therefore, a continual density-independent mortality could cause a decline in the population during the dry season that is counteracted by the onset of reproduction in the rainy season. Meanwhile, the same density-independent mortality could continue during the rainy season as well, but the population would still show a net increase during this time because of the reproduction occurring then. Thus, the steady-state pattern seen over the course of a year could be due to either density-dependent survivorship or fecundity (no one knows) but the life history predictions do point to fecundity as the more important, and data on seasonality in Puerto Rico are consistent with this interpretation.

While it is enheartening that the analysis has been taken this far, ideally one would like the foraging theory itself to be density-dependent, eliminating the need for a purely phenomenological treatment of density dependence using the fitted coefficient a of Equations 1.58 and 1.63.

1.7 Discussion

A synthetic theory for a West Indian *Anolis* lizard has been developed in this chapter, as summarized below:

- A foraging lizard is imagined continually to decide whether to ignore or to pursue prey that appear within its field of vision. In this scenario, an optimal foraging policy is to pursue each type of prey only as far as that prey type's "cutoff distance."

- The lizard's aim is to maximize its net energetic yield, and Equation 1.32 is the formula for the net energetic yield, E/T, with three prey. For the lizard, one needs e_w, the energetic cost of waiting; e_p, the energetic cost of pursuit; and v, the pursuit velocity. For the prey, one needs the net energy content of an insect from each prey class, e_i; the flightiness coefficient of an insect from each class, f_i; and the abundance of insects in each class, a_i.

- From Equation 1.32 one then predicts the foraging cutoff distance for each of the three prey types, r_i, by finding the values of r_i that maximize E/T.

- From the predicted r_i, further details, including the diet and the average prey size, can also be calculated.

- By assuming that the lizard has a certain foraging time each day, a daily yield from the foraging is then calculated.

- By converting this yield into body growth, a curve is generated that predicts how a lizard's size would increase as it ages if the lizard's energy throughout its life were allocated to growth.

- By assuming that the daily foraging yield could be allocated to reproduction instead of growth, and upon including data on mortality, the intrinsic rate of increase and the steady-state population size can be calculated as a function of the age at which the daily yield is switched from growth to reproduction.

- Finally, the age of first reproduction is calculated as the age for switching from growth to reproduction that maximizes the steady-state population size, on the assumption that density affects fecundity once maturity has been reached and does not affect survivorship or growth rate up to maturity.

Each step in developing this synthetic model involved a comparison with data on anoles. Data were available for all the coefficients except the flightiness coefficients, f_i, and the number of foraging hours per day averaged over a lizard's entire life. Guesses were offered for these later coefficients based on indirect evidence. Overall, the model appears to capture both qualitatively and quantitatively much about an anole's biology as an individual.

Furthermore, this chapter has shown that optimal foraging behavior as predicted by this model can be "learned," in the sense that this behavior can quickly be attained by a lizard that follows a simple and zoologically plausible "rule of thumb." No one has investigated whether a lizard does actually follow such a simple rule of thumb, but there have been investigations into how well lizards [51, 38, 268] and other animals [117] learn. These investigations point out that early learning studies of lizards were largely unsuccessful because workers used unimportant cues and reinforcers or conducted tests at environmental temperatures too low for the lizards to become active. Recent studies have shown that learning in lizards depends on temperature and that heat itself can be used as a reinforcer. Even more recently, field studies on anoles have shown that food is effective as a reinforcer in instrumental discrimination conditioned learning [330]. Thus, it is now clear that early studies greatly underestimated lizard intelligence.

The need for incorporating behavioral decision rules into foraging theory has been recognized since at least 1978 by Krebs [173], and learning has been increasingly included in some way in foraging theory since then [233, 132, 160, 164, 162, 163, 218, 219]. While the initial importance of a rule of thumb is to confer a behavioral plausibility to predictions of foraging theory, this feature may not be the most important ultimately. By placing foraging theory in the context of behavioral rules, one obtains a dynamic, rather than a static, picture of behavioral development. Figures 1.1, 1.10, 1.15, and 1.18 illustrate a dynamic approach to optimal foraging behavior, whereas traditional optimal foraging theory predicts only a static end point.

Many issues remain to be explored theoretically concerning rules of thumb. First, the role of a finite-time memory time could be investigated in relation to how well an animal's foraging behavior tracks temporal changes in the environment. Second, the use of Scheme permits innovation to be studied, wherein a lizard remembers not only what its energy yield and time commitments have been but also *how* it has gone about pursing prey. By allowing the lizard to vary its technique with each pursuit, one could model the development of complex ability: the packaging together of elementary abilities into a single composite ability. Third, the codevelopment of social behavior, such as the development of simple territoriality, could be modeled as asynchronous processes representing each lizard following its own rule of thumb. Then a social system would become manifest as the aggregate expression of the behaviorally interacting individuals. In Unix, for example, each forager could be spawned as an independent process (with the `fork()` system call), and then the collection of processes could access a common plane on which the insects appear. A social system might then emerge in this population.

Fourth, the sizes of feeding territories could be determined as the result of a game. The yield realized from these territories combined with the interindividual spacing that results would lead to an individual-based theory of density dependence and carrying capacity [83]. Fifth, this foraging theory could be used in a theory of body size in relation to male reproductive success [368]. Female territories are contained within male territories, and a male defends a territory both to feed in it and to contain females. As a male becomes bigger, it presumably can defend a larger territory, but its food needs increase too. Thus, the foraging theory and its rule of thumb introduced in this chapter should be viewed as a starting place, not an end in itself.

I find it hard not be jealous of the sentient lizard's intelligence. If lizards can learn to do the right thing, why can't humans? The answer, of course, is that if humans also followed a rule of thumb that leads to optimal behavior, then they too would learn to do the right thing. This issue is not as fanciful as it may seem. Managers often must seek to implement optimal policies. Yet managers rarely have the time or inclination to solve a dynamical optimization model to determine what the optimal policy is. But by following a rule of thumb, managers can discover during practice what an optimal course of action is. Thus, perhaps the sentient forager will serve as an example to motivate further research into rules of thumb in economics and decision theory, so that deans may aspire to become as smart as lizards.

Finally, we should have the temerity to inquire whether there is anything our sentient lizard cannot do. Because its computer program can expand dynamically as it runs, perhaps our sentient lizard has no limits. Perhaps our sentient lizard can prove theorems and tell right from wrong.

2

Invasion and coevolution

Economic development and human population growth are shrinking natural habitat and changing its geometry. From an airplane one easily sees disjointed fragments of forests and plains that were continuous stretches of natural habitat a few years ago.

The loss of natural habitat and its fragmented geometry imply that society is increasingly purchasing substitutes for ecosystem services it used to enjoy for free. Such services include control of insect pests; access to a diversity of evolutionarily tested drugs, spices, and foods; a source of living decoration and art; and space and scenery for recreation. Because economic development is often desirable, and certainly inevitable, it's useful to compare the economic benefit of development with the cost of lost services when deciding whether to proceed in particular instances. To make this comparison we must predict in advance how an ecosystem will respond to human modifications. We have to know how ecosystems work—what makes them tick, and what happens when we add or subtract pieces. But where will we get this knowledge? Where are some sample ecosystems whose workings can be exposed? This quest leads to islands. They are natural laboratories both for ecology and evolution, and their scientific value was first brought to general attention by Charles Darwin's discovery of evolutionary change during his visit to the Galapagos Islands.

Since the beginning of ecology as a science, and long before today's increasing environmental awareness, scholars have been interested in how ecosystems[1] work. The Caribbean anoles initially attracted ecological attention when patterns were discovered suggesting an inherent structure in ecosystem organization. The idea was first enunciated by G. E. Hutchinson [156] in 1959 who asked: "How much difference between two species at the same level [in the food web] is needed to prevent them from occupying the same niche?" The point to this question is that species are able to coexist if each occupies a different niche,[2] in some sense, whereas their coexistence is prohibited by competition for resources if they occupy the same niche, provided resources are in fact limiting and that predation and disturbance are low. To answer this question, Hutchinson followed Lack

---

[1]An ecological community—integrated set of interacting biological populations—has traditionally been distinguished from an ecosystem—community plus surrounding physical environmental. Today though, community studies, and even studies of single populations, are also likely to include purely physical processes, and the only distinction remaining between a community and ecosystem is in emphasis. Community studies tend to equate ecosystem components with specific biological populations while ecosystem studies tend to rely on aggregate variables such as biomass at various trophic levels.

[2]In ecological jargon, a niche is a species' occupation, and a habitat is its address.

[175] and Brown and Wilson [46] in using metric morphological characters as an index of niche differences between species. Hutchinson accumulated data on the bill lengths of coexisting birds and the skull length of coexisting mammals. For any pair of species, one species is usually larger than the other, and Hutchinson observed that: The ratio of the larger to the small form varies from 1.1 to 1.4, the mean ratio being 1.28 or roughly 1.3. This latter figure may tentatively be used as an indication of the kind of difference necessary to permit two species to co-occur in different niches but at the same level of a food web. The body sizes of the anoles throughout the eastern Caribbean appear to lend themselves to this interpretation and to offer an excellent model system in which to unravel the role of competition in determining community structure.

So, we begin with the pattern of body sizes through out the eastern Caribbean. We ask: Are there differences in the body sizes of coexisting anoles? Are anoles of the same body size forbidden from coexistence by competition? The eastern Caribbean has nearly 30 islands across which patterns in body size can be observed, whereas the Galapagos comprises less than 10. The Caribbean therefore offers a more detailed comparative picture of ecosystem structure than available in the early literature—a picture that reveals both a regularity in body sizes as well as some exceptions to the general pattern. We will then face the task of explaining both the regularities and the exceptions.

To explain body-size patterns, scenarios are developed for how communities of competing species can be assembled. Because these scenarios depend on whether anoles actually do compete for food, the evidence for interspecific competition between species of anoles is reviewed. Then the scenarios are tested against biogeographic, fossil, and phylogenetic data. To learn the answer, read on.

The pattern of body sizes is not the only biogeographic pattern revealed by eastern Caribbean anoles. Another pattern involves what may be called the "landscape" aspect of anole distributions within an island. On some islands one species peaks in abundance at the island's center while a second species peaks around the island's periphery near sea level. On other islands both species cover the entire island and increase and decrease in synchrony from place to place. The chapter therefore concludes with hypotheses to explain this difference in the landscape ecology of anoles and sets the stage for the still larger scale of the next chapter—the geologic origin of the Caribbean.

2.1 Biogeography of body size

2.1.1 Overview of the islands

Twenty-nine islands that dot the margin of the eastern Caribbean, each with a native anole species, are named in Table 2.1 in an approximately clockwise order beginning with St. Croix near Puerto Rico and ending with Curacao near Venezuela. Studies of island biogeography [202, 87, 59] have revealed two generalizations, both of which turn out *not* to apply to anoles on these islands. The number of species on an island is often an increasing function of the island's area—the "area effect"—and the number of species on an island, when corrected for island area, is often a decreasing function of distance from a source fauna—the "distance effect." Yet in the eastern Caribbean, the highest species diversity of anoles (two species) is on islands intermediate in size and in distance from the possible source locations of Puerto Rico or Venezuela. Neither the area nor distance effects are observed by this fauna. This foreshadows a deep role for plate tectonics in the biogeography

Table 2.1 Statistics for eastern Caribbean Islands.

Species	Area (km²)	Elevation (m)	Anoles
St. Croix	218	1087	1
Carrot Rock (BVI)	0.013	28	1
Anguilla	91	65	1
St. Martin	88	424	2
St. Barthelemy	21	302	1
Saba	13	857	1
St. Eustatius	31	598	2
St. Kitts	176	1315	2
Nevis	93	1093	2
Barbuda	161	34	2
Antigua	280	401	2
Redonda	1	304	1
Montserrat	85	913	1
Guadeloupe	1513	1464	1
Ilet-à-Kahouanne	0.2	69	1
La Désirade	18	278	1
Petite Terre	1	20	1
Marie Galante	93	200	1
Terre de Haut	3	318	1
Terre de Bas	6	284	1
Dominica	790	1443	1
Martinique	1116	1346	1
Barbados	430	339	1
St. Lucia	616	956	1
St. Vincent	344	1231	2
Grenada	311	838	2
La Blanquilla	47	60	1
Bonaire	246	243	1
Curaçao	448	372	1

of anoles, an effect of ancient history that is absent in birds, mammals, or insects, which are either more vagile or evolutionarily younger than anoles but which have been more prominent in the island biogeography literature than reptiles have been.

The anoles on the eastern Caribbean Islands span five lineages (cf. Chapter 3 for more detail). Beginning in the northeast and working around the islands in a clockwise order, the lineages are: (1) *A. acutus* from St. Croix is related to Puerto Rico anoles. (2) *A. ernestwilliamsi* is endemic to a tiny cay in the British Virgin Islands and is closely related to the widespread *A. cristatellus* of the Puerto Rican bank. (3) The anoles from the Anguilla Bank through Dominica belong to the bimaculatus group, which are distantly related to Puerto Rican anoles. (4) The anoles from Martinique through Bonaire belong to the roquet group, whose only known relative is the anole of a small island, Malpaleo, in the Pacific near Columbia. (5) *A. lineatus* on Curaçao is distantly related to Jamaican anoles.

At present, the most important of these lineages to remember are the bimaculatus group in the north and the roquet group in the south. Thus, the anoles from these islands indicate whether ecological generalizations transcend evolutionary lineages.

2.1.2 The one-species islands

The solitary size

The solitary anoles all have about the same size, with only one exception, as noted in Table 2.2. The statistic used for an anole species size is the maximum reported SVL. Other writings have used as statistics the head length of the largest one-third of the sample [306, 307] and the average jaw length [282], and the maximum reported SVL [383]. To achieve repeatability and wide participation of data sources, I have followed Williams [383] in using the maximum SVL, even though this statistic depends somewhat on collecting effort. A technicality, however, is that specimens shrink after preservation in ethanol, typically by about 9% [180]. Therefore by convention, the maximum SVL is defined as that measured on preserved specimens, *or* as measured on freshly collected specimens multiplied by 0.91 to compensate for the anticipated shrinkage. (Alternatively, the lengths of live animals can be approximated by dividing all the values in Table 2.2 by 0.91, giving an equivalent picture.) The data are from Lazell [180, 181] for the Lesser Antillean and Virgin Island anoles, Hummelinck [155] for La Blanquilla and Bonaire, and our own collections for Curaçao and for *A. marmoratus alliaceus* of Guadeloupe.

The body size on single-species islands (called the "solitary size") is, from 19 islands excluding Marie Galante, 77 mm for males and 55 mm for females. The "midsex" (average of male and female maximum size) is 66 mm (i.e., (77 + 55)/2). These and other summary statistics appear in Tables 2.5 and 2.6. The sexual dimorphism, expressed as the ratio of average male to average female maximum SVL is 1.39. On finer examination, the bimaculatus group is somewhat smaller than the roquet group, with a midsex of 64 mm and 73 mm, respectively[3] [383].

Marie Galante: An outlyer

A. ferreus of Marie Galante is an outlyer: Its males are far larger than solitary

---

[3] In calculating these averages, the nominate subspecies was used, *A. m. marmoratus* for Guadeloupe, *A. o. oculatus* for Dominica, and *A. r. roquet* for Martinique.

Table 2.2 Maximum snout-vent length on one-species islands.

Island	Species	Male	Female	M/F
St. Croix	*A. acutus*	65	48	1.35
Carrot Rock (BVI)	*A. ernestwilliamsi*	75	55	1.36
Anguilla Bank	*A. gingivinus*	72	53	1.36
Saba	*A. sabanus*	69	50	1.38
Redonda	*A. nubilus*	81	52	1.56
Montserrat	*A. lividus*	70	52	1.35
Guadeloupe	*A. m. marmoratus*	82	54	1.52
	A. m. alliaceus	79	57	1.39
	A. m. girafus	75	56	1.34
	A. m. setosus	69	51	1.35
	A. m. speciosus	73	52	1.40
	A. m. inornatus	65	50	1.30
Ilet-à-Kahouanne	*A. kahouannensis*	73	48	1.52
La Désirade	*A. desiradei*	80	53	1.51
Petite Terre	*A. chrysops*	66	48	1.38
Marie Galante	*A. ferreus*	119	65	1.83
Terre de Haut	*A. terraealtae*	80	54	1.48
Terre de Bas	*A. caryae*	75	55	1.36
Dominica	*A. o. oculatus*	79	58	1.36
	A. o. cabritensis	81	59	1.37
	A. o. montanus	96	64	1.50
	A. o. winstoni	86	64	1.34
Martinique	*A. r. roquet*	86	66	1.30
	A. r. summus	82	62	1.32
	A. r. majolgris	82	50	1.64
	A. r. zebrilus	82	63	1.30
	A. r. caracoli	79	58	1.36
	A. r. salinei	78	62	1.26
St. Lucia	*A. luciae*	91	63	1.44
Barbados	*A. extremus*	83	60	1.38
La Blanquilla	*A. blanquillanus*	85	65	1.31
Bonaire	*A. bonairensis*	75	60	1.25
Curaçao	*A. lineatus*	74	57	1.30

males and are typical of the larger species on a two-species island. Females of *A. ferreus*, while large, are not beyond the range of some other solitary females. The net result is a very large sexual dimorphism, with a male-to-female ratio of 1.83, which is about the ratio of the larger to the smaller species on the two-species islands. This dimorphism is also greater than that within the larger species of two-species islands and is the greatest dimorphism in the eastern Caribbean. The midsex for *A. ferreus* is 92 mm.

Subspeciation

The islands in the center of the Lesser Antillean arc (Guadeloupe, Dominica, and Martinique) contain single species that are divided into subspecies. A subspecies is a geographical race with a strikingly distinctive appearance. If specimens representing each subspecies were placed side by side, then someone familiar with lizards, but not specifically with these islands, would unhesitatingly suspect that each subspecies was a different species. Yet the subspecies completely intergrade with one another from the center of one subspecies' range to the center of another's. Thus, the subspecies nomenclature is a way of describing a dramatic geographical variation in phenotype.

Of the islands with two species, neither shows enough geographical variation to have led to the recognition of subspecies, although *A. aeneus* of Grenada comes the closest.

Clines in body size

Body size varies somewhat among habitats within an island.[4] The maximum male SVL on Guadeloupe varies among the subspecies from 65 mm to 82 mm, on Dominica from 79 mm to 96 mm, and on Martinique from 78 mm to 86 mm. Roughly speaking, the smallest anoles are found in xeric habitat, and the largest in middle-elevation forest. The maximum female size does not vary as much as the maximum male size. All in all, a high body size is associated with a large insect abundance, high rainfall, and warm temperature, and some plasticity can be detected as well [287]. Further studies of the geographical variation are presently underway in Dominica [203].

So, the key points to remember about the solitary anoles are that they have a body size in common of about 66 mm, within an island there is some variation in body size among habitats, and Marie Galante has an exceptionally large anole.

2.1.3 The two-species islands

Symmetrical displacement

The larger species on the islands with two species excluding St. Martin is much larger than the solitary size. Males usually exceed 100 mm in maximum SVL, and females exceed 65 mm in maximum SVL. The midsex for the large species of the two-species islands is 93 mm. All this is noted in Table 2.3. The anoles from the two-species islands are compared with the solitary anoles in Tables 2.4 and 2.5.

The smaller species on the islands with two species is less than or equal to the solitary size. If the smaller species on bimaculatus islands are compared

---

[4] Variation in phenotype along a geographical transect is called a "cline."

Table 2.3 Maximum snout-vent length on two-species islands.

Bank	Island	Species	Male	Female	M/F
		Large member of species pair			
Anguilla	Guana Cay	*A. gingivinus*	72	53	1.36
St. Kitts	St. Eustatius	*A. bimaculatus*	90	67	1.34
	St. Kitts	*A. bimaculatus*	114	69	1.65
	Nevis	*A. bimaculatus*	112	66	1.70
Antigua	Barbuda	*A. leachi*	113	70	1.61
	Antigua	*A. leachi*	111	69	1.61
St. Vincent	St. Vincent	*A. griseus*	127	86	1.48
Grenada	Tobago	*A. richardi*	125	77	1.62
		Small member of species pair			
Anguilla	St. Martin	*A. pogus*	50	42	1.19
St. Kitts	St. Eustatius	*A. schwartzi*	49	43	1.14
Antigua	Barbuda	*A. forresti*	52	45	1.16
	Antigua	*A. wattsi*	58	46	1.26
St. Vincent	St. Vincent	*A. trinitatis*	74	57	1.30
Grenada	Grenada	*A. aeneus*	77	55	1.40

with the bimaculatus solitary size, and if the smaller species of roquet islands are compared with the roquet solitary size, then one can say that all the smaller species of two-species islands are less than the corresponding solitary size. But if the smaller species on the two-species islands are compared with the grand mean of the solitary species, then one can say only that the smaller species of two-species islands are less than, or equal to, the solitary size.

Thus, not only are the species of two-species islands different from one another, by a factor of about 1.6 to 1.8, also they each tend to differ in opposite directions from the solitary size.

St. Martin: An outlyer

The Anguilla Bank is an outlyer among the two-species islands. *A. gingivinus*, ubiquitous across the entire bank from Sombrero to St. Barths, is a solitary-sized anole. Table 2.3 records its size on a cay just south of St. Martin, and it is no larger elsewhere, including where it co-occurs with another species. *A. pogus* is today found only in the hills surrounding Pic du Paradis in the center of St. Martin. However, it was collected in 1922 on Anguilla and was absent there by 1966 [180]. *A. pogus* has gone extinct on Anguilla between 1922 and 1966.[5] Moreover, *A. pogus* appears to have evolved a smaller size since it was first collected. The largest specimen of *A. pogus*, 50 mm SVL, was collected in 1883; this is the

---

[5]It is one of only two endemic populations of *Anolis* known to have done so in the eastern Caribbean, the other being the giant *A. roosevelti* of Culebra near Puerto Rico.

Table 2.4 Comparison of one- and two-species island maximum SVL.

	Average	Stand. Dev.	Stand. Err.	Number
2-species islands — smaller species				
Males-All	60.00	12.44	5.08	6
Females-All	48.00	6.39	2.61	6
M/F = 1.25				
Males-Bimac.	52.25	4.03	2.02	4
Females-Bimac.	44.00	1.83	0.91	4
M/F = 1.19				
Males-Roquet	75.50	2.12	1.50	2
Females-Roquet	56.00	1.41	1.00	2
M/F = 1.35				
1-species islands — solitary species				
Males-All	76.89	6.99	1.60	19
Females-All	55.32	5.51	1.26	19
M/F = 1.39				
Males-Bimac.	75.18	5.53	1.67	11
Females-Bimac.	52.46	2.98	0.90	11
M/F = 1.43				
Males-Roquet	84.00	5.83	2.60	5
Females-Roquet	62.80	2.77	1.24	5
M/F = 1.34				
2-species islands — larger species				
Males-All	113.14	12.08	4.56	7
Females-All	72.00	7.12	2.69	7
M/F = 1.57				
Males-Bimac.	108.00	10.12	4.52	5
Females-Bimac.	68.20	1.64	0.73	5
M/F = 1.58				
Males-Roquet	126.00	1.41	1.00	2
Females-Roquet	81.50	6.36	4.51	2
M/F = 1.55				

Table 2.5 Comparison of one- and two-species islands continued.

	Midsex average of maximum SVL		
	Small	Solitary	Large
All	54.00	66.11	92.57
Bimaculatus	48.12	63.82	88.10
Roquet	65.75	73.40	103.75
	Ratio of midsex averages		
	Solitary/Small	Large/solitary	Large/small
All	1.22	1.40	1.71
Bimaculatus	1.33	1.38	1.83
Roquet	1.12	1.40	1.58

specimen cited in Table 2.3. Today though, specimens this large are never found; also, *A. pogus* is conspicuously smaller, by about 10%, than its relatives on the St. Kitts bank (*A. schwartzi*) and the Antigua Bank (*A. forresti, A. wattsi*).

So, the Anguilla Bank is exceptional in that the larger anole has a solitary size instead of the very large size typical of the larger species on the other two-species island banks; the smaller species has gone extinct on Anguilla; and, where the smaller species still exists on St. Martin, it has apparently decreased in size during this century and is smaller than its relatives on the St. Kitts and Antigua Banks.

St. Eustatius: An intermediate

The larger anole on St. Eustatius is not as large as it is on nearby St. Kitts and Nevis. Most authors have lumped this island with St. Kitts and Nevis nonetheless, because St. Eustatius is smaller and dryer than either St. Kitts or Nevis, and smaller lizards might be expected there as a result. Alternatively, St. Eustatius could be regarded as a transient state "on the way to" the condition found today on St. Martin.

Recent changes in large species

The size quoted for *A. richardi* of Grenada is based on specimens from Tobago, a land bridge island near Venezuela to which *A. richardi* has been introduced. The species is evidently slightly larger there than in its source habitat on Grenada. Similarly, *A. leachi* was introduced to Bermuda sometime about 1940 from Antigua and has apparently increased slightly in body size on Bermuda [259]. Thus, the larger species of two-species islands are capable of evolving to be even larger than they are, and occasionally have done so when introduced to new habitats.

So, the key points to remember about the two-species islands are that the larger species is larger than the solitary size; the smaller species is less than or equal to the solitary size; and the Anguilla Bank is exceptional in having a ubiquitous solitary-sized larger species and a smaller species restricted to the central hills of St. Martin following its recent extinction on Anguilla.

Table 2.6 Maximum male SVL of largest and smallest species.

Island	Diversity	Smallest	Largest	L/S
Cuba	>35	38	191	5.03
Hispaniola	>35	38	175	4.61
Puerto Rico	11	40	137	3.43
Jamaica	7	57	124	2.18

2.1.4 Greater Antilles

The Greater Antilles possess species more extreme in size than those of the Lesser Antilles, as noted in Table 2.6 [383, 385]. The maximum SVL of males in the smallest species, generally called "twig anoles," is only about 40 mm, whereas the maximum SVL of males in the largest species, known as "giant anoles," reaches 190 mm. The ratio of large to small is therefore almost a factor of 5, with many species filling the space in between. Species of anoles are still being discovered in Hispaniola and Cuba. Not only do the Greater Antilles have anoles more extreme in size than any of the Lesser Antilles, but also some species occupy niches and have life styles qualitatively different from those in the Lesser Antilles. The Greater Antilles are like continents in the *Anolis* universe, and the islands of the Lesser Antilles might be hoped to shed light on how the Greater Antillean communities are put together.

The Greater Antilles offer a glimpse of a common blueprint from which the anole communities at many places are constructed. Williams [385] introduced definitions of "ecomorphs" as listed in Table 2.7, based on Rand's description [263, 264] of "structural niches" in the faunas of Puerto Rico and Jamaica (also cf. [314, 315]). Williams's idea is that a community in a sufficiently productive area, such as woods or forest, contains as many as six coexisting anole species distributed throughout the habitat from the canopy to the ground in characteristic places and feeding in characteristic ways. The canopy includes three species, a crown giant, twig dwarf, and trunk-crown anole. The later is exemplified by the common green anole, *A. carolinensis*, often seen in pet stores. These canopy anoles all differ in size. Then progressing down from the canopy to the ground, and alternating in size, are more forms culminating in a bush-grass anole. The trunk-ground anole is perhaps the most conspicuous anole to humans because it is usually very abundant and is easily seen near human eye level. While the data are far from precise, the lowland Puerto Rican and Hispaniolan fauna appear to follow this plan. The agreement is less clear in Jamaica and dubious in Cuba, although a recent analysis of the Greater Antillean communities emphasizes convergent aspects [195].

The species that belong to each ecomorph in Puerto Rico are listed in Table 2.8. Several species belong to the same ecomorph in some cases: three trunk-ground anoles and three bush-grass anoles. The most interesting fact about species of the same ecomorph is that they do not coexist locally; instead they occupy different regions of the island (cf. [137, 154, 144]). The order in the table when there are multiple species per ecomorph is from mesic to xeric.

The community structure of the Greater Antillean anoles resembles the structure of other vertebrate communities, notably the fruit pigeons of New Guinea

Table 2.7 Ecomorphs of *Anolis* lizards in the Greater Antilles.

Ecomorph	Size	Color	Modal perch	Body proportions
Crown giant	>100	Green	High canopy	Massive head often casqued
Twig dwarf	<50	Gray	Canopy, vines	Long head, short body and legs
Trunk-crown	>70	Green	Canopy, trunk	Long body, short legs
Trunk	<50	Varied	Central trunk	Short head and body
Trunk-ground	>60	Brown	Lower trunk	Large head, short stocky body,
Bush-grass	<50	White Stripe	Grass, bushes	Long head, tail Slender body

Table 2.8 Species of each ecomorph in Puerto Rico.

Ecomorph	Species	SVL
Crown giant	*A. cuvieri*	137
	A. roosevelti	157
Twig dwarf	*A. occultus*	40
Trunk-crown	*A. evermanni*	78
Trunk	*A. stratulus*	50
Trunk-ground	*A. gundlachi*	72
	A. cristatellus	74
	A. cooki	62
Bush-grass	*A. krugi*	55
	A. pulchellus	51
	A. poncensis	48

studied by Diamond [87, 88] and the rodents of the Great Basin and Sonoran deserts studied by Brown [44].

What then, is the relation of these large communities to the small communities of the Lesser Antilles? Are Lesser Antillean communities early stages in the development of a common blueprint for Greater Antillean communities?

2.2 Early hypotheses

2.2.1 The invasion scenario

Perhaps the simplist explanation that can be offered for how the *Anolis* communities of the eastern Caribbean have been organized is to combine Hutchinson's rule [156] with the body sizes in the eastern Caribbean, as follows:

First, we see if a two-species island can be assembled from solitary components. The maximum extreme within the solitary anoles (other than *A. ferreus* on Marie Galante) is *A. oculatus montanus*, whose midsex is 80 mm. The average midsex for the bimaculatus solitary anoles is 64 mm, leading to a ratio of 1.25. Therefore, a species with the size of *A. oculatus montanus* could probably coexist, but barely, with an average species having the solitary size. The situation improves somewhat if we compare *A. oculatus montanus* with the minimum extreme of the solitary forms *A. marmoratus inornatus*, whose midsex size is 58 mm, leading to a ratio of 1.38. So, by choosing source and destination sizes at the extremes of the range for solitary anoles, cross-invasions are possible according to Hutchinson's rule. In particular, a plausible scenario is for a large montane anole to be transported on vegetation that is carried down a river after a storm and that washes up against the typical xeric habitat of sea level where the anoles are relatively small.

Second, if we have one two-species island, can we get any more two-species islands? By Hutchinson's rule both the larger species and the smaller species of the existing two-species islands (other than those from St. Martin) could move laterally to a single-species neighbor, leading to an additional two-species island.

Third, once we have a two-species island, the large species continues to stay large, and the small continues to stay small because of competition. That is, the size differences are acquired in allopatry but preserved in sympatry.

According to this scenario, once the initial two-species island is formed others will follow by cross-invasion. If so, the large species of the two-species islands will be related to one another, the small species to one another, and both groups will trace to solitary anoles. This is in fact true in the bimaculatus group, though not in the roquet group (cf. phylogenetic tree in Chapter 3).

Hutchinson's rule offers another prediction. The ratio of large to small SVL's for the two-species islands is about 1.6 to 1.8, which is much larger than 1.3. If Hutchinson is correct, then the anoles of most two-species islands are much more different than they need to be to coexist. Indeed, perhaps the anoles on most two-species islands don't compete at all—except on St. Martin where the ratio of body sizes is about 1.2.

The essence of an invasion scenario is that the species differences that today permit coexistence were acquired in allopatry as adaptations to different ancestral habitats. Because the differences exceeded Hutchinson's ratio, invasions could take place. After an invasion, with two populations now in a common habitat, the species differences are maintained because neither species can evolve into the other's niche because it is already occupied. However, a major alternative to this type of hypothesis postulates that the species differences were acquired in sympatry (by character displacement).

2.2.2 The character-displacement scenario

The idea of "character displacement" traces to the book *Darwin's Finches* by Lack, in 1947, wherein differences in body size (and bill dimensions) of finch species were postulated to permit their coexistence, much as body size does in anoles. What is unusual, though, is that Lack could observe the same species both as a solitary population and as a member of a two-species community. In Figure 17 of *Darwin's Finches* (p. 82) two islands each have a *solitary* ground-finch with a bill depth of about 9 to 10 mm. On one island the solitary species is *Geospiza fortis* and on the other island the finch is *G. fuliginosa*. But on three islands these same two species coexist, and there *G. fuliginosa* has a bill depth of about 8 mm, while *G. fortis* has a bill depth of about 12 to 13 mm. Thus, when it co-occurs with a competitor, each species has diverged from the size it has when it is solitary. Indeed, each of the species has diverged in opposite directions from the solitary size. This situation invites the interpretation that the two solitary-sized species diverged from each other when they came in contact, and if so, the differences in body size they now exhibit were acquired in sympatry when the contact was initiated. This phenomenon of divergent evolution has been called "character displacement" by Brown and Wilson [46] who provided more examples, and called general attention to its theoretical importance.[6]

In 1972 Williams [383] sketched a hypothesis for how the fauna of small islands—the anoles of the Lesser Antilles—can be imagined as early stages in the assembly of a complex fauna—the anoles of Puerto Rico—using the idea of character displacement. Williams wrote: Assume the invasion of the domain [island] of a solitary anole by successful propagules of another solitary anole. The two are initially *ex hypothese* similar in size. They compete for food and other resources. The classic expectation is divergence (character displacement) in order to avoid or lessen the effects of competition. Among the possible kinds of divergence will be size divergence, which has been shown to correlate with utilization of different sizes of food. ... Our model and our prediction then is that, as a result of interspecies competition, one species will become larger and the other smaller.

The character displacement model would seem to explain the regular one-species islands and the regular two-species islands, but what about the outlyers of Marie Galante and St. Martin? Williams called attention to these outlyers, and offered ways to reconcile these exceptional situations. For Marie Galante the simplest escape is to hypothesize that a small species has very recently become extinct there, perhaps as a result somehow of human activities. There is no evidence of this, and human habitat use there seems roughly the same as that on other islands. Alternatively, Williams suggested that recurring invasions to Marie Galante from Guadeloupe, each of which fails, would nonetheless "nudge" *A. ferreus* to be larger than expected of solitary anoles. Still Williams was not happy with either of these suggestions and wrote that "The Marie Galante anole is on any hypothesis a special case; there are no genuine parallels."

---

[6]The phrase "character displacement" sometimes is used to refer solely to the *pattern* of species traits being displaced in opposite directions in sympatry when compared with their allopatric states, and sometimes to the *process* of divergent coevolution that leads to the pattern. Here we use character displacement to refer to the process of divergent evolution.

Concerning St. Martin, Williams emphasized that *A. gingivinus* contacts *A. pogus* in a small area. The data at that time were consistent with viewing *A. pogus* as physiologically limited to damp shady places such as small ravines. Therefore, *A. gingivinus* could be regarded as having a solitary size because it actually is solitary throughout most of its range. In contrast, *A. pogus* had to evolve downwards in response to competition from *A. gingivinus* because it co-occurs with *A. gingivinus* throughout *its* entire range. Thus, the key is an asymmetry in the range of overlaps: *A. gingivinus* hardly notices *A. pogus*, whereas *A. pogus* sees *A. gingivinus* everywhere.

It was not appreciated at the time that St. Martin is every bit as troublesome as Marie Galante. The range of *A. pogus* was underestimated; it is actually found throughout the hills of St. Martin, not only in ravines. About one-fourth to one-third of the area of St. Martin has both species, an area of contact presumably large enough to have a reciprocal evolutionary effect. In fact, no suggestions were offered as to *why* the St. Martin condition exists at all.

Finally, Williams extended the idea of invasion followed by character displacement to "construct" a complex fauna. Empirically, 40 mm is about as small as an anole species apparently can get, and 200 mm is about as large. So, Williams postulated that anole sizes are bounded between this floor and ceiling. Next, Williams imagined that successive invasions followed by reciprocal evolutionary readjustments led to body size ratios of about 1.5 to 2.0 between adjacent species, culminating in a maximum of five species in a community with sizes of, say, 40 mm, 60 mm, 90 mm, 135 mm, and 200 mm. Thereafter, Williams suggested that any additional species added to the island would have to live in different regions of the island, such as desert or high montane habitat. Williams concluded by showing that the fauna of Puerto Rico could be assembled formally according to this rule. Whether actual invasions and coevolutionary episodes occurred in this way was, of course, not known. Still, it was a great advance to exhibit a plan by which a complex fauna could be generated from the extension of phenomena displayed in simple faunas.

Today then, we have to evaluate two propositions: Are the body size differences produced in sympatry by character displacement, or are the differences acquired in allopatry? Are small faunas really precursors to large faunas? Does the path lead from up from 1 to *n* species? Do rules of sequential assembly really underly community structure?

Because both Hutchinson's rule and character displacement depend on competition as the process that underlies community structure, do different species of anoles really compete with one another for food, and does the strength of competition really depend on difference in body size? There's not much point in proceeding further with hypotheses, tests, and so forth, until this question is answered.

2.3   Evidence for competition

During the mid 1970's to mid-1980's, both descriptive and experimental evidence accumulated that insular anoles compete for food, so that there is now general consensus on the reality of competition in insular anoles. Readers not familiar with ecology often fail to appreciate the significance of this consensus. Even casual observation reveals plants competing for light, a garden's need for water, and the flush of green in a newly fertilized lawn. One is inclined therefore to assume that competition for resources is the norm, and, indeed, survival of the

Table 2.9 Abundance and body size on Bonaire.

Site	Insects	*A. bonairensis*	M SVL	F SVL
Well	463 (49)	13.0 (2.6)	62.4 (1.8)	50.3 (1.2)
Quadrat 1	314 (32)	4.7 (0.4)	54.1 (2.8)	49.1 (0.9)
Quadrat 2	324 (21)	5.0 (1.0)	57.2 (3.1)	50.4 (1.0)
Quadrat 3	292 (27)	5.0 (1.3)	55.7 (3.8)	50.9 (0.9)

fittest is popularly interpreted as a competition for resources. But for animals, and perhaps many plants as well, the existence of competition for limiting resources is far from obvious. One alternative is that that predation, disease, and disturbances from fire, hurricanes, and so forth keep abundances low enough that resources are not limiting. Another alternative is that populations are unregulated by any type of ecological interaction; they may be undergoing a slow random walk between bust and boom conditions, culminating eventually in extinction, but not before, on the average, budding off new species to take their place. Thus, for animals at least, survival of the fittest may have little to do with competition for actual resources such as food or space but instead may pertain to escape from predation, thermoregulation in severe physical environments, and so forth. Nonetheless, for insular anoles, competition for food *does* seem to be an extremely important, often the most important, ecological consideration underlying the structure of these ecological communities.

What should *not* be used as evidence for competition is biogeographic pattern, especially patterns related to body size. The pattern is what has to be explained— it is not the explanation. Observing a size ratio, such as 1.3, between coexisting species does not entail that competition is present. The theory goes in the other logical direction. Hutchinson's claim is that *if* two competing species coexist, *then* the species must differ by a ratio of about 1.3.

This section reviews data that have led to the consensus on food limitation and competition for insular anoles. Moreover, information on the strength and dimensionality of the competition are pointed out. Then in the next section, analysis of the hypotheses for community assembly is resumed.

2.3.1 Correlations with food supply

Anoles are hungry, and they put on weight when given live insect bait [186, 329, 328]. Licht [185] wrote of Puerto Rican anoles that "virtually all adult lizards are constantly alert and ready to accept food at all times of day during the summer breeding season. These lizards are rarely satiated; in fact, stomachs are usually nearly empty." Thus, individual anoles are food limited.

But is population abundance also determined by food quantity? Table 2.9 shows the abundance of anoles from four sites on Bonaire.[7] The units are lizards per 100 m^2, insects per m^2 per 12h, and SVL in mm. Bonaire is an arid single-species island. Three sites were adjacent replicates in scrub and show the same abundance and average body size of anoles. A fourth site was near a pond and

---

[7]Unpublished data taken with G. Gorman during August 17–24, 1975.

Table 2.10 Abundance on St. Martin.

Site	Insects	A. pogus	A. gingivinus
Boundary (scrub)	1415 (404)	0	50.1 (2.2)
Naked Boy Mt (scrub)	953 (104)	0	43.2 (3.4)
Well (woods)	215 (88)	0	42.4 (1.9)
Point Blanc (scrub)	96 (25)	0	6.4 (0.5)
Medium Pic (woods)	289 (108)	17.7 (0.7)	10.2 (2.0)
St. Peter Mt (woods)	260 (65)	16.9 (0.8)	19.7 (1.4)
Pic du Paradis (woods)	164 (37)	10.5 (0.7)	8.4 (2.6)

damp stream bed. The anoles from the fourth site were both more numerous and larger in body size.

On St. Martin, too, the anoles are more abundant where there are more insects. Sea-level sites have only *A. gingivinus*, while sites in the central hills of the island have both *A. gingivinus* and *A. pogus*, as noted in Table 2.10 from a study during August 23 to September 7, 1977. The units are lizards per 100 m^2 and insects per m^2 per 12h. Among the sites where *A. gingivinus* is solitary, its abundance is correlated with insect abundance, although the relation is nonlinear. The abundance appears to plateau at about 0.5 lizard per m^2 (i.e. 50 per 100 m^2), when censused in late August.

Thus, correlations with food supply suggest both that individual lizards as well as the abundance of solitary populations are limited by food. But is there any indication of an interaction *between* species from abundance data?

2.3.2   St. Martin and St. Eustatius

Distributional evidence—St. Martin

Even if two populations of anoles are each limited by food, they still may not compete, because each population may mostly consume food that the other does not use. If so, no interspecific competition exists, and the abundance of one species does not affect the abundance of the other. But on St. Martin the abundance of *A. gingivinus* is consistently lower at the sites where it co-occurs with *A. pogus* than at the sites where it is the solitary anole (excepting the very arid Point Blank), as noted in Table 2.10. This is indirect evidence of interspecific competition.

Furthermore, *A. gingivinus* changes its perch when it co-occurs with *A. pogus*: It moves higher into the vegetation, and *A. pogus* takes over the lower perch positions, as noted in Table 2.11. The units are fractions of observations in each height class and total number of observations. This is further circumstantial evidence of interspecific competition.

Finally, St. Martin, and to a lesser extent St. Eustatius, are the only two-species islands wherein the smaller species has a restricted distribution. On Barbuda, Antigua, St. Kitts, and Nevis, both species are found throughout the islands, whereas on St. Martin and St. Eustatius the species are closer in body size and do not co-

Table 2.11 Perch distribution on St. Martin.

Species	State	Gnd	0–0.46	0.46–0.9	0.9–1.8	>1.8	N
A. gingivinus	Solitary	0.13	0.27	0.16	0.30	0.13	419
A. gingivinus	Co-occur	0.05	0.12	0.12	0.33	0.40	111
A. pogus	Co-occur	0.15	0.25	0.23	0.28	0.08	250

occur throughout the entire island. Instead, the smaller species predominates in the woods toward the center of these islands, while the larger species predominates in the open habitat at sea level. This difference in the islandwide distributions that suggests competition on St. Martin and to a lesser extent on St. Eustatius restricts the smaller species to the central hills where conditions are cooler. The smaller species of St. Martin and St. Eustatius have a lower tolerance for heat stress [291], as discussed further below.

The circumstantial evidence of competition is important because it provides a motivation for experimental studies. Experiments to detect competition in systems lacking any distributional or other indirect evidence of competition seem likely to fail, a priori. However, the circumstantial evidence cannot stand alone, because it is consistent with an alternative hypothesis that *A. gingivinus* is simply an arid-adapted anole and that *A. pogus* is simply a mesic-adapted anole. If so, the hills are marginal habitat for *A. gingivinus* and the sea level marginal habitat for *A. pogus*, independently of each other. The perch shift, for example, could represent *A. gingivinus* moving into the canopy to be closer to the sun for thermoregulatory needs and need not represent the result of a present-day ecological interaction with *A. pogus*. Hence, field experiments were carried out.

Introductions to Anguillita

Anguillita is a tiny cay at the western tip of Anguilla, and like the other cays of the Anguilla Bank, it contains a resident population of *A. gingivinus*. Four propagules of *A. pogus* from nearby St. Martin were introduced to Anguillita during experiments between August 1979 to October 1981 [294, 291]. In August 1979, a site of 560 m^2 was surveyed surrounding a large "sea grape" tree (*Cocoloba uvifera*). It was found to have 196 anoles, as indicated on the first row of Table 2.12. The units are the total number of lizards in the site. Then 103 individuals of *A. pogus* was introduced. Eight months later, in May 1980, the site was recensused and found to have only about 15 of the *A. pogus* remaining, while the resident population of *A. gingivinus* was unchanged in size. Meanwhile, in March 1980 another site was surveyed having 260 m^2 and 116 resident anoles, approximately the same density as the first site. At this site about 35% of the resident anoles were collected, and then 55 individuals of *A. pogus* were introduced. Seven months later about 16 of these introduced anoles remained, and the population of *A. gingivinus* regained its size of 116. This survivorship is almost twice that of the first trial in which none of the resident *A. gingivinus* were removed. Also, a hatchling of *A. pogus* was observed, indicating that reproduction was occurring. This basic design was then repeated, as also listed in Table 2.12. Removing about half the resident *A.*

Table 2.12 Introduction to Anguillita.

Experiment	State	Date	*A. gingivinus*	*A. pogus*
SK	Original	August 1979	196.3 (7.2)	0
558.6 m^2	Start	August 1979	196.3 (7.2)	103
	End	May 1980	204.6 (9.3)	14.5 (7.0)
MK	Original	March 1980	115.6 (8.9)	0
262.6 m^2	Start	March 1980	74.6 (8.9)	55
	End	October 1980	116.4 (5.4)	15.5 (1.7)
SK2	Original	March 1981	194.1 (14.5)	10.3 (7.6)
558.6 m^2	Start	March 1981	92.1 (14.5)	110.3 (7.6)
	End	October 1981	153.5 (3.3)	37.3 (2.4)
LU	Original	March 1981	103.9 (5.1)	0
295.9 m^2	Start	March 1981	103.9 (5.1)	48
	End	October 1981	127.2 (5.5)	7.0 (1.6)

gingivinus nearly doubles the six-month survivorship of the introduced *A. pogus*, as noted in Table 2.13 where the units are the fraction that survived six months.

This experiment is evidence of strong present-day competition by *A. gingivinus* against *A. pogus*, and indicates that *A. pogus* cannot be successfully introduced into scrubby sea-level habitat already occupied by *A. gingivinus*. If *A. gingivinus* is removed, *A. pogus* can survive in sea-level habitat. There is no indication that *A. pogus* is physiologically incapable of living and reproducing in this habitat. Instead, *A. pogus* is competitively excluded from sea-level habitat by *A. gingivinus*.

Enclosures on St. Martin

The Anguillita experiments demonstrated a competitive effect of *A. gingivinus* against *A. pogus*. Indirect evidence of the reciprocal interaction comes from the distributional data mentioned earlier, in which *A. gingivinus* has an upwards perch shift and a lower abundance where it co-occurs with *A. pogus* relative to where it is solitary. The next set of experiments provide direct evidence of this reciprocal interaction and also provides direct evidence that the strength of competition between populations depends on difference in average body size.

These experiments were conducted using a system of fences and cleared vegetation that contained the animals introduced within and excluded the animals from the surrounding habitat [241]. The technique consists of a wire mesh fence that is constructed with the lower edge buried to about 10 cm, to prevent lizards from tunneling under the fence. The mesh is small enough to prevent even hatchlings from poking their way through. The fence is about 1.2 m high and is topped with a flat piece of polypropylene that overhangs the fence on each side. Anoles cannot climb on this material, and therefore cannot climb over the fence. The enclosure

Table 2.13 Effect of *A. gingivinus* on survivorship of *A. pogus*.

Experiment	6-month survivorship of *A. pogus*	% removal of *A. gingivinus*
SK₁	0.230	0
LU	0.193	0
MK	0.337	35.5
SK₂	0.396	52.6

Table 2.14 Effect of *A. pogus* on *A. gingivinus* in enclosures.

Indicator	GW1	GW2	G1	G2
F growth	.0093(.0021)	.0095(.0024)	.0181(.0025)	.0166(.0023)
M growth	.0084(.0018)	.0075(.0020)	.0154(.0020)	.0131(.0019)
F max SVL	49.8(3.2)	49.5(3.7)	52.3(1.5)	52.1(2.0)
M max SVL	62.8(5.4)	60.9(4.2)	63.9(2.2)	63.9(3.0)
Egg Vol.	70.5(17.8)	45.5(11.5)	200.3(23.4)	139(20.9)
Stomach Vol.	36.8(11.7)	36.0(13.8)	108.9(29.1)	72.4(15.2)
Perch Ht.	.89(.05)	.87(.06)	.38(.01)	.49(.03)

is 12 m × 12 m overall, and 2 m at each side of the fence is cleared of vegetation. This clearing extends to the canopy so that lizards cannot cross into the enclosure by jumping from tree to tree, or from a tree on the outside to the ground on the inside, or vice versa. With little maintenance the experiment can be carried out for six months; after about two years the plastic becomes brittle and needs replacing.

The experimental design consisted of four enclosures on St. Martin and on St. Eustatius [238, 240]. Table 2.14 presents the results for St. Martin with units as described below. During January 1981 the four enclosures on St. Martin were each stocked with 60 *A. gingivinus*, and two were also stocked with an additional 100 *A. pogus*. The total abundance of anoles in the control enclosures was 60 and in the treatment enclosures was 160. Given that the area of the enclosure is 144 m², these densities bracket the nominal natural value of one anole per m². At monthly intervals from January 1981 through May 1981, the *A. gingivinus* individuals were noosed and their lengths and weights measured. At the end of the experiment most of the *A. gingivinus* individuals were collected for analysis of stomach contents and reproductive condition.

The first two columns of Table 2.14 describe the treatment enclosures containing both *A. gingivinus* and *A. pogus* (GW1 and GW2), and the last two columns describe the control enclosures containing solely *A. gingivinus* (G1 and G2). The growth data were obtained by fitting the monthly changes in weight to a logistic equation [316, 100]. The growth rate is thus the juvenile growth rate, and the

Table 2.15 Effect of *A. schwartzi* on *A. bimaculatus* in enclosures.

Indicator	BW1	BW2	B1	B2
F Growth	.0058(.0021)	.0095(.0023)	.0078(.0019)	.0078(.0018)
M Growth	.0060(.0016)	.0060(.0016)	.0075(.0021)	.0064(.0018)
F max SVL	65.8(6.0)	62.6(2.6)	63.6(4.5)	62.0(3.0)
M max SVL	85.1(6.5)	84.1(6.1)	80.1(5.4)	83.9(7.2)
Egg Vol.	35.7(20.0)	58.6(28.1)	41.0(15.1)	52.0(22.0)
Stomach Vol.	113.5(23.2)	54.7(11.9)	111.2(30.1)	65.1(17.7)
Perch Ht.	2.04(0.12)	2.07(0.14)	2.07(0.15)	1.90(0.13)

projected terminal size is also tabulated.[8] The units of growth are mm per d, projected maximum SVL is in mm, egg volume is in mm^3 per female anole, stomach contents are mm^3 per anole, and perch height is in m. The standard error is in parentheses. The data show that both males and females were growing at a slower rate in the treatment enclosures than in the control enclosures, although they were tending toward the same maximum SVL, the amount of egg material in females from the treatment was less than in females from the controls; the stomach content was lower in the treatment than in the controls; and the perch height was higher in the treatment than in the controls (as already seen in the comparative data on perch height distribution). The five indicators of competitive effects are each statistically significant. All five, taken together, indicate a very strong effect of the *A. pogus* individuals on *A. gingivinus*.

These exclosure experiments on St. Martin and the introductions to Anguillita provide experimental evidence of strong reciprocal competition between the anoles of the Anguilla Bank. Recall that the size ratio of the anoles on St. Martin is about 1.2, a value less than Hutchinson's critical value of 1.3. The strong competition observed on St. Martin tends to confirm Hutchinson's rule that competition precludes coexistence between species whose body size ratio is appreciably less than 1.3.

Enclosures on St. Eustatius

On St. Eustatius the midsex of the larger species is 78.5 mm, and for the smaller is 46 mm, leading to a ratio of 1.7. This ratio is generously greater than Hutchinson's critical value of 1.3, suggesting that competition may be weak or nonexistent between the anoles of St. Eustatius. Table 2.15 presents results from the enclosure experiment carried out on St. Eustatius, using the same design as on St. Martin. Here the four enclosures were stocked with 60 *A. bimaculatus* and two were also stocked with an additional 100 *A. schwartzi*. As on St. Martin, the experiment was also run from January through May 1981. The first two columns indicate the treatment enclosures (BW1 and BW2), and the second two columns indicate the controls (B1 and B2). The result is that there is no result. The treatments do not

---

[8]In Chapter 1 a Bertalanffy model was preferred, as illustrated in Figures 1.22 and 1.23, and in the future these data should be reanalyzed on the basis of a Bertalanffy model rather than a logistic model.

Table 2.16 Effect of *A. bimaculatus* on *A. schwartzi* in enclosures.

Indicator	BN1	BN2	A1	A2
F growth	.0129(.0010)		.0131(.0008)	
M growth	.0117(.0006)		.0134(.0005)	
F max SVL	40.7(.6)		41.2(.5)	
M max SVL	47.4(.6)		47.5(.4)	
Egg Vol.	39.3(16.7)	46.8(15.7)	66.1(16.0)	41.7(12.8)
Stomach Vol.	17.3(2.2)	19.6(2.9)	19.9(2.7)	26.3(5.1)
Perch Ht.	.13(.01)	.22(.02)	.76(.08)	.98(.10)

differ from the controls, indicating that the presence of 100 additional anoles in the treatment enclosures has no effect on the *A. bimaculatus* there. This finding implies that the 100 *A. schwartzi* do not consume food or other resources that the 60 *A. bimaculatus* are using, and therefore their presence does not matter. Thus *A. schwartzi* does not have a strong (or even any) competitive effect on *A. bimaculatus*.

The significance of the comparison between the St. Martin and St. Eustatius experiments is that the two St. Martin species are very close in body size, while the two St. Eustatius species are fairly different in size. Because the strength of competition was much higher on St. Martin than on St. Eustatius, strength of competition evidently does depend on difference in body size—the greater the difference the less the competition. Moreover, Hutchinson's value of 1.3 is confirmed. Further evidence that the strength of competition depends on difference in body size has since appeared for *Parus* birds in Europe [3] and introduced birds in Hawaii [227].

When this St. Martin/St. Eustatius comparison was completed, data were lacking concerning reciprocal competition (or the lack thereof) on St. Eustatius. Hence, the enclosures on St. Eustatius were renovated and used in 1982 to ascertain whether *A. bimaculatus* exerts a competitive influence on *A. schwartzi* [298]. Four enclosures were stocked with 80 *A. schwartzi* apiece, and two had an additional 40 *A. bimaculatus* added. The experiment was run from February through May 1982. Table 2.16 presents the results. The first two columns pertain to the treatment enclosures with both *A. schwartzi* and *A. bimaculatus* (BN1 and BN2) and the second two columns to the control enclosures with *A. schwartzi* alone (A1 and A2).[9] The table shows the treatment did not have an effect. Therefore, neither species had a detectable effect on the other on St. Eustatius, while both species did have a detectable effect on the other on St. Martin.

Niche dimensionality

The St. Eustatius setup was used to explore a further issue sometimes called the "dimensionality of niche space." The question is whether the separation in perch positions between the anoles on St. Eustatius is a distinct axis along which to divide up limiting resources (such as space suitable for territories) independent

---

[9]For the growth data, BN1 and BN2 were combined, and A1 and A2 were combined.

Table 2.17 Effect of perch-lowered *A. bimaculatus* on *A. schwartzi*.

Indicator	BL1	BL2
F growth	.0116(.0008)	
M growth	.0120(.0007)	
F max SVL	40.6(.5)	
M max SVL	47.4(.7)	
Egg Vol.	50.2(21.9)	13.2(3.7)
Stomach Vol.	16.3(1.9)	21.7(3.5)
Perch Ht.	.21(.02)	.12(.02)

from the axis of body size, which can be thought of as a way to divide up the food resources with respect to prey size. Alternatively, the difference in perch height could be a secondary consequence of the primary difference in body size and not an independent axis at all. That is, perhaps lizards perch above each other *because* they have different body sizes. The large *A. bimaculatus* may perch above the small *A. schwartzi* because that is how it gets the best vantage point for seeing the relatively rare and large prey that it is willing to chase at great distances. If so, the niche space is really one-dimensional, with the difference in perch height being derivative on the difference in body size.

To examine this issue, the perch positions of *A. bimaculatus* were experimentally lowered, so that *A. bimaculatus* were forced to perch closer to *A. schwartzi* than normal. This was achieved by putting polypropylene collars around the trees at a height of 1.5 m so that the higher part of the trees and canopy were inaccessible to *A. bimaculatus*. Table 2.17 shows the results (BL1 and BL2).[10] Perhaps surprisingly, the perch-lowered *A. bimaculatus* had no larger effect on *A. schwartzi* than *A. bimaculatus* at normal perch height, which is no effect at all. In fact, the only measurable effect of the perch lowering was a deleterious effect on *A. bimaculatus* itself, as seen in Table 2.18. As a result of using lower perches, the *A. bimaculatus* caught less food and grew more slowly. If perch height were a distinct niche axis, then an effect on *A. schwartzi* on *A. bimaculatus* should have been detected. The conclusion, then, is that the perch height differences observed in the bimaculatus two-species islands are simply a reflection of the difference in body size of the species: Each species is perching where it can best forage for the prey sizes appropriate to its body size. Hence, these communities appear to have an essentially one-dimensional niche space. It should be emphasized though that this conclusion is not true for the roquet two-species islands, nor of course, for the Greater Antilles or Central America.

Sensitivity to temperature

Some species in the northern Lesser Antilles on the two-species island differ in their sensitivity to heat stress, the smaller species being the more sensitive. On St. Eustatius, individuals of both *A. bimaculatus* and *A. schwartzi* have the same

---

[10]Growth data of BL1 and BL2 combined.

Table 2.18   Effect of perch-lowering on *A. bimaculatus*.

Indicator	BL1	BL2	BN1	BN2
F growth	.0025(.0006)		.0051(.0008)	
M growth	.0029(.0005)		.0037(.0005)	
F max SVL	66.0(4.2)		61.2(1.2)	
M max SVL	83.8(5.5)		84.8(4.2)	
Egg Vol.	25.0(23.4)	7.1(7.1)	72.1(32.7)	49.2(32.5)
Stomach Vol.	20.3(5.4)	53.8(16.4)	82.5(17.7)	57.6(9.4)
Perch Ht.	1.00(.09)	1.00(.07)	2.44(.17)	2.13(.17)

body temperatures when collected at a site where both occur [289, Figure 16.3c]. Similarly, on St. Martin [291, Figure 10.3] individuals of both *A. gingivinus* and *A. pogus* have the same body temperature at the Pic du Paradis site where both occur. Nonetheless, an indication of differing sensitivity to heat stress can be found by holding a lizard in the open sun, allowing it to heat up. When it first begins to open its mouth to pant, its body temperature is noticed and then it is released. The distribution of "panting temperatures" obtained in this way shows that *A. schwartzi* is more sensitive to heat stress than *A. bimaculatus* because it begins to pant at a lower temperature [289, Figure 16.3a], and that *A. pogus* is more sensitive to heat stress than *A. gingivinus* because it too pants at a lower temperature [291, Figure 10.4]. As already mentioned, the abundance of both these small species peaks in the wetter, cooler sites above 300 m above sea level. This differing sensitivity of heat stress presumably has an impact on competition. At hot and xeric sites, the total number of minutes per day when the smaller species can be active is presumably less than that available to the larger species. These measurements have not been carried out on Barbuda where the two species seem equally at home in the most xeric of habitats, so the generality and phylogenetic affinity of these observations are not known. As mentioned later, these findings definitely do not generalize to the southern Lesser Antilles, where the opposite relationship appears to occur.

2.3.3   Other introductions

The record of recently introduced anoles supplies more evidence of interspecific competition. To aid the biological control of the fruit fly, *Ceratitis capitata*, in 1905 the director of agriculture of Bermuda liberated 71 individuals of *A. grahami* in public gardens [393]. *A. grahami* is a medium-sized lizard; it was collected from the Kingston area of Jamaica. By 1940 *A. grahami* was abundantly distributed throughout Bermuda. About 1940 the large anole from Antigua, *A. leachi*, first appeared. The species is now established in the center of Bermuda (Warwick area), primarily in woods. The spread of this large anole has taken place on territory already occupied by the medium-sized anole. Finally, the solitary-sized anole, *A. extremus*, from Barbados was noted in 1953 at the western end of Bermuda (Somerset and Ireland areas) where one may observe that it still has a patchy and limited distribution today.

Hispaniola provides two well-documented instances of "enclaves" of introduced anoles; these typically form when natural habitat is cleared and replaced with plantings. The Cuban green anole, *A. porcatus*, has been known since 1970 from a few city blocks in Santo Domingo at the site of former trade fairs [384, 255]. Its range has remained static through 1977 and is surrounded by the native green anole, *A. chlorocyanus*. The Puerto Rican trunk-ground anole, *A. cristatellus*, has been known since 1956 to be abundant in gardens in the port city of La Romana in the Dominican Republic [384, 114]. The town is the site of a sugar mill constructed by a Puerto Rico–based company. The range of the anole has been static for over 20 years and is surrounded by the native trunk-ground anole, *A. cybotes*.

Experimental enclaves have also apparently been produced in the Bahamas on tiny cays that were already inhabited by an anole [319, 320]. In contrast, introductions to empty cays large enough to support anoles produced a population explosion of the introduced species, although these reports must be viewed as preliminary.

Initial evidence suggested that introduced small anoles do not form enclaves when surrounded by larger anoles, but become extinct instead. A population of *A. wattsi* known from the botanical garden in St. Lucia was reported to be extinct [369, 384]. However, J. Losos has kindly sent me a copy of a letter by D. Corke reporting that *A. wattsi* is "firmly established in and around Castries" and has confirmed this report with his own observations (cf. [77]). In contrast, in the experimental introductions to Anguillita mentioned previously, *A. pogus* did fail to establish in the presence of a resident *A. gingivinus* [294, 291].

The only small anole clearly successful as an invader is the trunk-ground anole, *A. sagrei*, of Cuba that has invaded Jamaica [382], Grand Cayman island [197], and Florida [75, 166, 384, 391, 300, 366]. *A. sagrei* is quite small in comparison to the trunk-ground anoles of Puerto Rico and Hispaniola. It is both smaller and occupies more open habitat than the trunk-ground anole of Jamaica, *A. lineatopus* [264]. It is also smaller and uses hotter microsites than *A. conspersus* of Grand Cayman [17, 197]. The fact that *A. sagrei* differs from its competitors in *both* body size and body temperature may account for its success as an invader.

A recent review of 22 *Anolis* introductions by Losos, Marks, and Schoener [197] concluded that "Both documented instances of failure of an introduction involved ecologically similar species" and that "all six of the introductions that have been successful (i.e., the introduced species is abundant and widely distributed) have been cases in which the introduced species is ecologically dissimilar to the resident." They also noted that for anole communities, the composition of the resident community was more important than the total number of species in it in predicting invasion success, whereas in larger communities, such as bird communities, colonization success can be inversely correlated with number of species in the native community [57, 58]. Also, some direct evidence of competition between two similarly sized trunk-ground anoles of Puerto Rico has appeared for *Anolis cooki* and *Anolis cristatellus* [161, 207].

All in all, the record of introductions seems to suggest that, provided the niche separation is primarily along a body-size axis only and not also along a microclimate axis, a large anole can become established in the presence of a smaller anole; an anole introduced to newly opened habitat can form a virtually static enclave in the range of another anole with similar size and habits; and the evidence is presently ambiguous as to whether a small anole is destined to rapid extinction if introduced in the presence of a larger anole.

2.3.4 Competition summarized

The data on competition establishes that food resources are limiting and that competition can occur between populations of anoles. The data confirm Hutchinson's criterion that competition sets in strongly at a body-size ratio of about 1.3 (for a linear dimension such as SVL) and that competition is weak or absent between species that differ by a ratio much greater than 1.3. The data also suggest that the bimaculatus two-species communities have a one-dimensional niche space and that the smaller species on some of the bimaculatus two-species islands is more sensitive to heat stress than the larger species.

2.4 Theory of faunal assembly

Now that we are assured that interspecific competition between anoles actually occurs in nature, we can attempt to refine the hypotheses to explain the biogeographic patterns in body size in the Lesser Antilles and how the simple communities are assembled.

We have considered an invasion scenario based on Hutchinson's rule, which makes no mention of coevolution between the species after an invasion has taken place. We have also considered the character displacement scenario, which assumes an invasion has somehow taken place prior to the start of divergent evolution. Because the invasion scenario does not treat coevolution after an invasion, and the character displacement does not treat the invasion before coevolution begins, a synthetic theory that includes both processes would seem helpful. Such a theory that includes both invasion and coevolution as alternating stages in the overall process of faunal assembly will now be presented. It is an idea model, not a model that is designed with anoles, or any other particular organism in mind.[11] After having seen what this model has to offer, we will return to the data and evaluate the alternative scenarios as best we can.

2.4.1 Premises

We start with a niche axis, x. This is interpreted as the logarithm of the lizard's length. A species with a particular body size has a "position" on the niche axis that is the logarithm of its snout-vent length. Another species with a different SVL has a different position on the axis. If species-1 has position x_1 and species-2 has position x_2, then the "distance" between the species on the axis is $x_1 - x_2$. Because the axis is in log units, a distance measured along it corresponds to a ratio of body lengths. Thus, Hutchinson's *ratio* of 1.3 corresponds to some *distance* along the niche axis that separates the species. This idea of a niche space was introduced by MacArthur and Levins in 1967 [200] and developed further by May in the early 1970's [214].

Next we need a model for the population dynamics of competing populations. The Lotka-Volterra competition equations—the equations found in all ecology textbooks—may suffice. Let's try them out and see what happens. These equations have as their parameters, r_i, the intrinsic rate of increase for species-i; K_i, the carrying capacity for species-i; and $\alpha_{i,j}$, the competition coefficient for the effect

---

[11]This type of theory has however, been tailored to whiptail lizards of Baha California in Mexico [56] and Darwins finches in the Galapagos [305].

of an individual of species-j against an individual of species-i. The idea then is to combine these equations with the concept of a niche axis.

We imagine that each position on the axis has a corresponding K. For example, if x is the log of the SVL in some population, then $K(x)$ denotes the carrying capacity of a population having that SVL. $K(x)$ is called the "carrying capacity function." One may guess that this function is more or less bell-shaped. The peak represents the optimum body size (the solitary size), and the bell shape corresponds to a log-normal distribution of K with respect to the SVL (i.e., a normal distribution with respect to the log of SVL). The most important parameter in the carrying capacity function describes the width of bell, σ_k^2, which can be thought of as the variance of the carrying capacity function. Moreover, it's also customary to translate the origin for x to lie under the peak of K, so that a negative x refers to a body size smaller than the optimum, and a positive x to a body size larger than the optimum. The other parameter in $K(x)$ is K_o, which is simply the height of the bell; the value of this parameter is arbitrary and can be used to scale the population sizes to convenient values.

The intrinsic rate of increase, r, also depends on the niche position. We take $r(x)$ also to be a bell-shaped curve that is simply a constant times $K(x)$, following Fenchel and Christiansen [111]. The idea is that both the intrinsic rate of increase and the equilibrium population size depend somehow on resource quantity and therefore should increase and decrease together, in accord with simple birth-death models. The value for r_o, the height of $r(x)$ can be used to scale the time. By using a value of 1, a fast invasion can take place.[12] Because $r(x)$ is the same shape as $K(x)$, its width is the same.

Last is another bell-shaped function to describe how the competition depends on the difference in niche positions. $\alpha(x_1 - x_2)$ indicates the competition coefficient of an individual at position x_2 against an individual at x_1. The competition coefficient indicates the effect of an *individual* at position x_2 against an individual at position x_1.[13] The reciprocal competition coefficient is $\alpha(x_2 - x_1)$. The competition function depends on the *difference* in niche positions, $x_1 - x_2$. Symmetrical competition (all reciprocal competition coefficients equal) is modeled with a bell-shaped curve centered at 0. α equals 1 at this peak and drops off to zero in both tails to indicate that highly dissimilar body sizes do not compete in either direction. The key parameter is σ_α^2, which determines the competition function's width. This width is typically interpreted as the population's "niche width."[14] Here, the competition width parameter, σ_α^2, is conventionally set equal to 1 (by scaling the x's, and everything else, accordingly), so that the only free width parameter in the theory is σ_k^2. The variables are the population sizes, N_1 and N_2 for invasions, and the niche positions, x_1 and x_2, for coevolution.

---

[12]One should avoid making r_o too high so that the model doesn't oscillate.

[13]It does not indicate competitive superiority of one *population* against another, for that depends also on the K's.

[14]The niche width can be partitioned into the sum of a "between-phenotype component" (BPC) and a "within-phenotype component" (WPC) [281]. The BPC describes how much of the population's overall variance in resource use is caused by differences among the individuals, and the WPC describes the variance in resource use of an average individual.

2.4.2 Invasion

invade.scm simulates an invasion. The program simply iterates the Lotka-Volterra equations,

$$\Delta N_1 = r_1(K_1 - N_1 - \alpha_{1,2}N_2)N_1/K_1 \qquad (2.1)$$

$$\Delta N_2 = r_2(K_2 - N_2 - \alpha_{2,1}N_1)N_2/K_2 \qquad (2.2)$$

where the parameters, r_i, K_i, and $\alpha_{i,j}$ are computed from the various bell-shaped functions given the niche position of each species. One assumes the resident, species-1, has its niche position at the peak of the bell-shaped curve: This is at $x_1 = 0$. Furthermore, the resident is assumed to be fully established on the island, so that its abundance N_1 equals $K(x_1)$. Then an invading propagule is introduced. Its niche position is x_2 and its initial abundance is rare, so N_2 equals some small initial value, say 1. Then by running the program one sees how N_1 and N_2 change through time, which reveals whether an invasion can take place, and if so, how long it takes. There are two milestones in an invasion—the time for the propagule to reach a significant size, say 1000, and the equilibrium population size that is ultimately realized by both the invader and the resident.

Figure 2.1 illustrates both the time to establish and the equilibrium population sizes as a function of the invader's niche position, for the case where σ_k^2 is 2. Notice that the best sizes for an invader to have are -1.7 and 1.7. These represent two scenarios, respectively: invasion by a smaller anole, and by a larger anole, against a resident whose size is 0. Any other size leads to a longer time for establishment. Indeed, an invader sufficiently similar to the resident takes an eternity to establish. Also notice that while ± 1.7 are the best places to invade, nearby positions are feasible too. Indeed, any place between 1 and 3, and between -1 and -3, seem fine; only positions very near the resident, or very far out in the tails, are effectively impossible. Therefore, while ecological competition sets constraints on where an invader can enter, the precise place on the niche axis where an invasion takes place is determined by what sizes of invaders are actually available within reach of the island.

2.4.3 Coevolution

The coevolution starts after the invasion is completed. coevolve.scm models the change in niche position of two coevolving competitors. The equations assume that body size is inherited as a quantitative character [289]. Here's the idea. Each species' niche position evolves in the direction that leads to an increase in mean fitness in that species for that generation. That is, if $W_1(x)$ denotes how the fitness in species-1 depends on niche position, then the slope of this function, $dW_1(x)/dx$, tells which direction leads to higher fitness within species-1. If this slope is positive, selection favors an increase in niche position of species-1, and if negative, a decrease in niche position. And similarly for $W_2(x_2)$. As in other quantitative genetic models, the speed of the evolution produced by this selection depends on the heritability and variability of body size in the progeny of matings. The fitness functions for both species, W_1 and W_2, themselves change each generation and must be recomputed each generation to take account of the new body sizes and population abundances that have resulted from the previous generation's evolution.

To use coevolve.scm, the initial niche positions of the resident and invader are given. The population sizes are assumed to be the equilibrium sizes for the

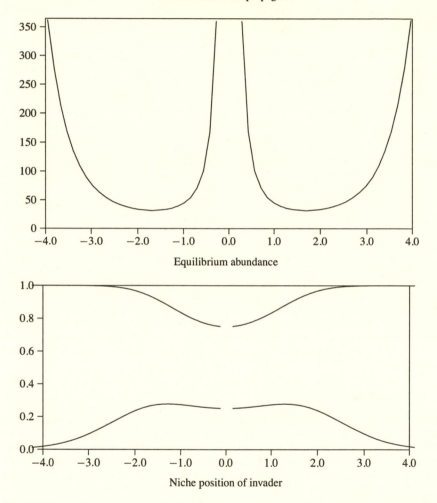

Fig. 2.1 Top: Time steps to establish propagule at an abundance of 1000 starting from an abundance of 1 as function of the position of the invader. Bottom: Equilibrium abundance ($/1,000,000$) of both species as a function of the position of the invader. Curve for resident at top of graph with a dip in the middle; curve for invader at bottom of graph with peaks at both sides of the middle. Position of resident is 0.0, and $\sigma_k^2 = 2.0$.

given niche positions. Then at each time step, new niche positions are computed, and the population sizes are assumed to catch up with the new niche positions that have just evolved. Eventually, a coevolutionary equilibrium may be attained.

Here's where the equations for `coevolve.scm` come from (also cf. [176, 334, 283, 335, 110, 289, 45, 364]. The basic formula from quantitative genetics for the change each generation in the mean value of a character is

$$\Delta \bar{x} = h^2(\overline{x_w} - \bar{x}) \tag{2.3}$$

where h^2 is the heritability of the character, $\overline{x_w}$ is the mean after selection, and $\bar{x}$ is the mean before selection. With weak selection, $\overline{x_w}$ works out to be approximately

$$\overline{x_w} = \bar{x} + \frac{\sigma^2}{W(\bar{x})} \frac{dW(\bar{x})}{dx} \tag{2.4}$$

where $W(\bar{x})$ is the fitness as a function of the character evaluated at the mean character, $dW(\bar{x})/dx$ is the derivative of the fitness function evaluated at the mean character, and σ^2 is the variance of the character in the population.

The conceptual equivalent of the Hardy-Weinberg law for a quantitative character is that the population variance, σ^2, approaches a constant, independent of the mean, $\bar{x}$, as a result of random mating and genetic segregation at the loci that determine the trait. Random mating tends to reduce population variation of the character, while genetic segregation tends to increase variation. These two processes balance one another, leading to a steady-state population variance estimated as $\sigma_L^2/(1 - h^4/2)$ where σ_L^2 is the average variance of the offspring produced by each mating, and h^4 is the heritability squared[15] [283, see Equation 9.12]. Putting all this together leads to an equation for evolutionary change in the mean value of the character as a result of natural selection,

$$\Delta x = \frac{s}{W(x)} \frac{dW(x)}{dx} \tag{2.5}$$

where s is a parameter that we'll call the evolutionary speed parameter, which is based on the genetic characteristics of the trait. (Its formula is noted below.) From now on we'll just write x instead of $\bar{x}$ even though the mean value of x is intended. For simple density-independent stabilizing selection, this formula shows that the mean value of the character climbs the fitness function and equilibrates where the fitness is maximized.

The quantative-genetic equivalent of the Hardy-Weinberg law is important because it disallows a single species from spreading out to fill the entire niche space by itself. Instead, the formula for Δx pertains solely to evolutionary changes in the mean value of x. The population variance for the character, σ, remains relatively constant through evolutionary time even though the mean is changing. Evolutionary change in σ requires modifying the underlying genetic system, a process that presumably can occur only in a population that occupies a stable habitat for a very long time.

---

[15] This quantitative-genetic equivalent of the Hardy-Weinberg law assumes that the population reproduces sexually. On the advantage of sexual reproduction, cf. [286].

To use this model of evolutionary change to study coevolution, we need a fitness function that represents the interspecific competition. With two species we have

$$W_1(x, x_1, x_2, N_1, N_2) = 1 + r(x) - \frac{r(x)}{K(x)}\alpha(x, x_1)N_1$$

$$-\frac{r(x)}{K(x)}\alpha(x, x_2)N_2 \qquad (2.6)$$

$$W_2(x, x_2, x_1, N_2, N_1) = 1 + r(x) - \frac{r(x)}{K(x)}\alpha(x, x_2)N_2$$

$$-\frac{r(x)}{K(x)}\alpha(x, x_1)N_1 \qquad (2.7)$$

This first expression refers to the fitness of an individual in species-1 at niche position x, given that species-1's mean position is x_1, species-2's mean position is is x_2, and the abundances of the species-1 and species-2 are N_1 and N_2, respectively, and similarly for the second expression. These fitness functions, based on the Lotka-Volterra format, are density-dependent and include both interspecific and intraspecific frequency dependence. For species-1, the derivative of W_1 has to be taken with respect to x, with everything else held constant, and then evaluated at $x = x_1$. Similarly for species-2, the derivative of W_2 is taken with respect to x, with everything else held constant, and evaluated at $x = x_2$.

In summary, then, the equations for how the niche positions of the two co-evolving species change through time is modeled by iterating the following two simultaneous equations starting with the initial condition that represents the niche positions following the successful invasion by the second species,

$$\Delta x_1 = \frac{s_1}{W_1(x_1, x_1, x_2, N_1, N_2)} \left.\frac{\partial W_1(x, x_1, x_2, N_1, N_2)}{\partial x}\right|_{x=x_1} \qquad (2.8)$$

$$\Delta x_2 = \frac{s_2}{W_2(x_2, x_2, x_1, N_2, N_1)} \left.\frac{\partial W_2(x, x_2, x_1, N_2, N_1)}{\partial x}\right|_{x=x_2} \qquad (2.9)$$

where s_i is the evolutionary speed parameter for species-i ($i = 1, 2$) interpreted as $\sigma_{L_i}^2/(1 - h_i^4/2)$.

Parallel evolution

Figure 2.2 illustrates four coevolutionary scenarios, again assuming the resident is located at $x_1 = 0$ and $\sigma_k^2 = 2$. The top panel refers to invaders larger than the resident, and the bottom panel to invaders smaller than the resident. Recall that if the invader is larger than the resident, then the best position for an invasion is 1.7. After such an invasion, the resident coevolutionarily shifts down to -0.5, and the invader shifts down to 0.5, as illustrated in Figure 2.2 (top, solid curve). Thus, *parallel evolution* and not divergent evolution has occurred. If the invader comes in at the other most likely size for establishment, -1.7, then both species evolve in parallel upwards, equilibrating at the same endpoint, -0.5 and 0.5, as illustrated in Figure 2.2 (bottom, solid curve). Invasion from below leads to a symmetric counterpart to invasion from above, and both scenarios involve parallel evolution as the coevolutionary equilibrium is attained.

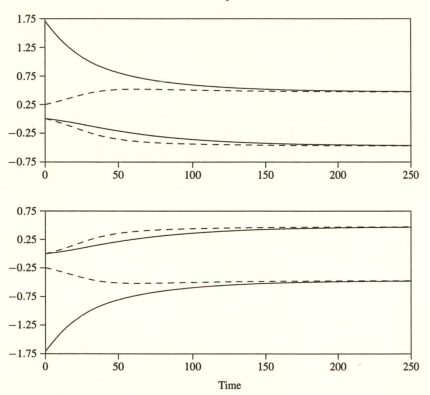

Fig. 2.2 Coevolving niche positions of resident and invader in arbitrary time units. Top and bottom graphs are symmetric equivalents: In the top, invader is larger than the resident; in the bottom, invader is smaller than resident. Solid curves for the invader begin from the most likely position for an invasion and show parallel evolution to the coevolutionary equilibrium. Dashed curves for the invader begin from a position improbably close to the resident and show character displacement during the approach to coevolutionary equilibrium. Curves for the resident always begin at solitary position, solid curves show coevolution with the most likely invader, and dashed curves show coevolution with an improbably close invader. The same coevolutionary equilibrium is attained regardless of invader's position. `vark` is 2, `evolutionspeed` 0.1

Character displacement

To obtain the character displacement scenario, the invader must be within 0.5 on either side of 0, i.e, closer to the resident at its time of invasion than it is at coevolutionary equilibrium. An invader at 0.25, for example, evolves upwards to 0.5, while the resident evolves downwards to -0.5, in accord with the traditional character displacement scenario, as illustrated in Figure 2.2 (top, dashed curve). The symmetrical counterpart of this, invasion as -0.25, is illustrated in Figure 2.2 (bottom, dashed curve). Notice though, that invasion at 0.5, or any closer, is difficult according to Figure 2.1. Therefore, the classical character displacement scenario is unlikely, though not impossible according to this model.

Taxon loop

It would seem that competition between anoles is not symmetric, because a large animal takes more food away from a smaller animal than vice versa. Also, interference competition would seem to favor a larger individual over a smaller one. To model this possibility a nonsymmetric competition function has been considered,

$$\alpha(x_1, x_2) = e^{\sigma_\alpha^2 \kappa^2/2} e^{-((x_1-x_2)+\sigma_\alpha^2 \kappa)^2/(2\sigma_\alpha^2)} \tag{2.10}$$

(cf. [283, p. 532]). This function is a bell-shaped curve that is displaced to the left and has a peak greater than 1, if the parameter κ is greater than 0. If the competitive asymmetry favors a large individual over a small one ($\kappa > 0$), then a solitary species comes to equilibrium not at $x = 0$, but at $x = \kappa\sigma_k^2$. That is, with asymmetry the solitary size is larger than the size at which $K(x)$ peaks [283, Equation 24.72].

invade.scm illustrates what this type of asymmetry can do. Suppose $\sigma_k^2 = 1.1$, and $\kappa = 1$. Because σ_k^2 is close to σ_α^2, there is barely room for two species. The resident's position as a solitary species is at 1.1, that is, at $\kappa\sigma_k^2$. The best sizes to invade turn out to be -0.9 and 2.1. Again we observe that invasion at nearly the *same* size as the resident remains difficult; this is difficult with or without asymmetry.

coevolve.scm shows that the scenario following this invasion is quite sensitive to the size of the species that does the invading and to the evolutionary speed parameter that includes the genetic information. As an example, suppose the invader is larger than the resident and has the size of 2.1, which is the best size for invasion from above. If the evolutionary speed parameter is ≥ 5.2, then during coevolution the resident becomes extinct. However, other scenarios are possible, including attaining a coevolutionary equilibrium with the resident at -0.2 and the invader at 1.08. Depending on the size of the invader, the trajectories leading to this equilibrium can involve parallel evolution up, parallel evolution down, divergent, or even convergent evolution. Thus, lots of invasion/coevolution scenarios seem to be possible according to traditional competition theory using the asymmetric competition function above. Of course, not all scenarios are equally likely, and the one feature that jumps out as fairly common is the presence of parallel evolution, either up or down, following an invasion.

The particular scenario in which a large invader excludes a medium-sized resident during coevolution leads to a "taxon loop." Once the resident goes extinct, the island returns to its previous one-species state and is ready to undergo another loop. Each loop is specifically triggered by the arrival of a large invader. Only one such invader may ever arrive, and if more than one does, the time between arrivals may be long and irregular. Also, perhaps no large invader ever arrives at a sufficiently remote island.

Passage to three species

The possibility of a two-species island being a stage to the buildup of a still larger community could be investigated next. One supposes the two resident species are at their coevolutionary equilibrium sizes and that the invader is rare. Then one tries all possible sizes for an invader and calculates the time needed to increase from a low abundance to some milestone abundance representing successful establishment. Examples appear in [299, 94]. The time for a third species to invade is generally longer than for the second to invade.

Effects of habitat change

The habitat for anoles in the Lesser Antilles has changed. Sea-level rise since the last ice age has reduced the area of all the islands, and human activities, especially since European colonization, have diminished the area of natural woodland and generally aridified the habitat. Such habitat changes might affect one species more than another. For example, higher temperatures could negatively affect the smaller species while not affecting the larger species on the Anguilla and St. Kitts Banks.

As an example, suppose the two anoles are at coevolutionary equilibrium and that some habitat change then affects the smaller species by lowering its K by some factor. Intuitively, the result is that the smaller species becomes less abundant than before and therefore makes less of a coevolutionary impact on the larger species. The larger species therefore approaches closer to the solitary size, and the smaller species evolves a still smaller size.

2.4.4 Hypotheses summarized

We began this modeling section with two scenarios, and two new scenarios have emerged. First, the invasion scenario postulates that the extreme sizes of the solitary populations can cross-invade, leading to a two-species island. One two-species island can then beget another. This scenario omits any reference to coevolution. Second, the character displacement postulates that two solitary species somehow meet and then diverge in body size, producing a two-species island. This scenario omits any reference to how two solitary-sized species could co-occur with one another to begin with. Third, a new scenario to emerge from the synthetic model is parallel evolution. A solitary island is initially invaded by species quite different from the resident. The resident then evolves away from the invader, and the invader evolves into the niche space just vacated by the resident. Both species thus coevolve in parallel, either down or up, depending on whether the invader was larger or smaller than the resident. Fourth, if there is asymmetrical competition favoring large species over small, then a large species that invades may evolve quickly enough to drive the smaller resident to extinction, restoring the system to one species, which then evolves to the position of the original resident (a taxon loop).

From a theoretical standpoint, these hypotheses are not equally plausible. The pure invasion scenario is plausible provided invasions happen quickly enough that the time between invasions is shorter than the time for coevolution. Character displacement is somewhat implausible, because the competition that is postulated to drive the divergent evolution will also inhibit an invasion, whereas parallel evolution is more robust theoretically. The taxon loop is theoretically the least probable. Variants on all these scenarios can be generated to accommodate the possible addition of a third species and to allow for habitat change.

Now let's see how these hypotheses stack up against historical data.

2.5 Historical evidence

Historical evidence concerning the biogeography of the Lesser Antillean anoles comes from fossils, archeological digs, and phylogenetic analysis.

2.5.1 Fossil anoles

Fossil anoles from the Tertiary have been located from Mexico and the Dominican Republic [179, 277, 252], as discussed further in the next chapter. All other fossils of anoles are latest Pleistocene to Holocene. Cave deposits, presumably left by the extinct barn owl, *Tyto cavaticus*, have been examined from the Bahamas, Jamaica, Hispaniola, Puerto Rico, Barbuda, and Antigua [107, 108, 109, 139, 256, 257, 258, 357]. Also, we discuss below data from our excavations on Anguilla [290].

The cave deposits reveal the present-day herpetofauna of the northern Lesser Antilles to be almost as diverse as the former fauna. Wholesale extinctions have *not* occurred since the Pleistocene, in contrast with insular birds [235], except until the recently introduced mongoose (*Herpestes*) began endangering ground lizards (*Ameiva* and *Iguana*) and snakes (*Alsophis*). The only clear extinction since the Pleistocene consists of populations of *Leiocephalus*, a ground lizard estimated at 150- to 200-mm snout-vent length, which have become extinct on Puerto Rico, the Antigua Bank, and, as our data show, also the Anguilla Bank. Its strongly tricuspid teeth are similar to the largely herbivorous lizard, *Dipsosaurus dorsalis* [157].

The Pleistocene cave deposits show that the larger anoles of the eastern Caribbean have become smaller. On the Antigua Bank, two separate excavations [107, 357] show that *A. leachi*, the larger of the two anoles today on the Antigua Bank, was still larger in the past; it has become smaller since the late Pleistocene. Also, the largest anoles from Puerto Rico have become smaller since the Pleistocene [256].

2.5.2 Excavations on Anguilla

Cave deposits are often found as isolated clumps of bones, discouraging the reconstruction of temporal sequence. Also, coarse screens have typically been used, and information on the smaller species is scanty. To obtain a quantitative record with improved stratigraphic resolution, a fissure in a cliff bordering Katouche Canyon on Anguilla was excavated [290]. Nests of the American kestrel (*Falco sparvarius*) are near the fissure, and these birds are the probable agents of fossil deposition.

Anolis pogus was last collected on Anguilla in 1922 [180]; only *A. gingivinus* remains today. Figure 2.3 shows the lengths for all *Anolis* dentaries; the top layer presumably contains only jaws from *A. gingivinus*, while the next layer contains jaws of *A. gingivinus* and *A. pogus* combined. The bottom scale is the snout-vent length corresponding to the dentary length from a regression using recent material. Table 2.19 presents the descriptive statistics for each layer and for pooled data from 0.4 to 1.4 ft and 1.7 to 4.1 ft.

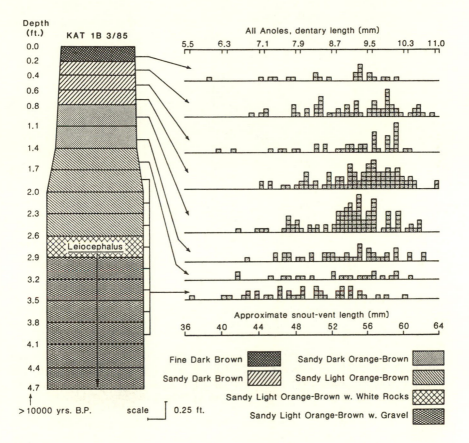

Fig. 2.3 Katouche Excavation on Anguilla. The profile of strata are at left, and histograms of jaw lengths at each stratum appear at right. Jaw length increases between the 0.2- and 0.4-ft strata; number of bones increases at about 1.4 ft. Charcoal fragments at the base of the excavation were dated as exceeding 10 Ka. Remains of *Leiocephalus* were found from 2.5 ft and below.

The excavation extended nearly 5 ft in depth. Charcoal fragments at the bottom were dated as greater than 10 Ka[16] (Washington State University, Radiocarbon Dating Laboratory, Sample #3196, Reported on Sep. 9, 1985).

Table 2.19 Size of *Anolis* dentaries from Katouche Excavation.

Depth (ft)	Num	Len (mm)	SE (mm)	SD (mm)
0.2	22	85.5	2.3	10.8
0.4	63	91.0	1.2	9.9
0.6	42	90.3	1.6	10.3
0.8	92	91.8	0.9	8.7
1.1	96	89.6	0.9	8.7
1.4	43	89.0	1.5	9.7
1.7	23	86.9	2.3	11.1
2.0	7	85.3	2.0	5.2
2.3	6	78.3	5.3	12.9
2.6	12	78.5	3.4	11.8
2.9	5	77.8	5.1	11.5
3.2	4	82.8	7.8	15.5
3.5	3	83.3	2.3	4.0
3.8	1	71		
4.1	2	87.5	8.5	12.0
4.4	0			
4.7	0			
0.4-1.4	336	90.5	0.5	9.2
1.7-4.1	63	82.9	1.4	11.1

The main features are: (1) The jaws are significantly smaller in the top layer than in the next layer (P < 0.02). (2) Numerous relatively large jaws extend from the second layer to about 1.5 ft. (3) Below 1.5 ft the jaws are fewer and smaller (P < 0.001). (4) Fragments from the jaws of *Leiocephalus* begin at about 3 ft and extend to the bottom.

An interpretation of Figure 2.3 is: (1) *A. gingivinus* has become about 10% smaller since the extinction of *A. pogus*. (2) *A. gingivinus* entered the island about 3000 to 4000 years ago. This interpretation is consistent with the appearance of both larger and more numerous jaws at the 1.5-ft level. (3) *Leiocephalus* became extinct about 5000 to 6000 years ago. (4) The original *A. pogus* was about 10% larger than it now is on St. Martin.

The first appearance of anole fossils at 4-ft depth should not be interpreted as the entry of *A. pogus*, because decomposition of fossils could explain the absence of anole jaws sufficiently deep in the profile; the number of all fossils declines toward the bottom of the excavation. In contrast, the absence of large jaws in the layers between 1.5 ft and 4 ft cannot be an artifact of fossil decomposition because

---

[16]Ka means 1000 years before the present.

larger jaws that preserve better are absent, while smaller jaws are present. A caveat is that the largest jaws coincide with the largest sample sizes. The interpretation above would be an artifact if some mechanism of deposition were more active during the time interval between the second layer and the 1.5-ft layer, resulting in the collection of exceptionally large specimens combined with the selective preservation of those large specimens. Still, there is no reason to suppose that this caveat is in fact true.

The evidence of reduction in size for the larger lizard on Anguilla, presented in Figure 2.3, is limited to the difference between the surface stratum and the one immediately below. This reduction in size would slightly predate, or coincide with, the extinction of the smaller anole on Anguilla during the 1920's. There is no evidence of change in size for the larger anole during the preceding 3000 to 4000 years, extending back to the time when the larger lizard appears to have been introduced. This fact, if genuine, would suggest that a small anole (say 50, mm in length) can precariously coexist with a solitary-sized anole (say, 65 mm in length). The small anole drops out only when the habitat is destroyed, which is the coup de grace given its strong competition with a larger anole. Still, the smaller anole does seem to have become somewhat smaller during that time interval (even though the larger anole did not change), because the smaller anole was larger at the time its competitor was introduced than it now is in its only remaining location on St. Martin.

2.5.3 Early humans

Archeologists have located remains of Stone age culture in the northern islands (St. Kitts and Antigua) dating to approximately 3 Ka [295]. Before this the Lesser Antilles may have been uninhabited. The arrival of Amerindians occurred about 2 Ka with the spread of the Arawaks from the Orinoco delta. Thus, Stone-Age man could have introduced *A. gingivinus* to the Anguilla Bank from Saba (where its closest relative lives), because lithic man may have entered the Lesser Antilles at about the time the Anguilla excavation suggests that *A. gingivinus* appeared and, at least today, man brings about many more introductions of *Anolis* than would occur by natural means. The cleanly separated distribution of bimaculatus and roquet groups across the island arc, however, argues against a major role for human introductions in the zoogeography of *Anolis*.

The extinction of *Leiocephalus* appears to predate the arrival of man. The sea level for 14 Ka was 40 m lower, by 8 Ka it was 15 m lower, and by 6 Ka it had achieved its current height [138]. Hence, the extinction of *Leiocephalus* approximately coincides with the culmination of the sea-level increase that reduced the exposed area of the Anguilla Bank to its present size.

Pregill [259] has suggested that the size change of the large species on Barbuda since the Pleistocene reflects a response to human habitat modification and to the introduction of animals associated with man. Human activities, especially wood cutting, renders the habitat more open. It is imagined that the food supply is reduced because of desertification (resulting in slower growth rates for lizards) and that lizards receive more predation pressure (implying that they do not live long enough to obtain the size they did prior to human activity). There is no reason to believe, however, that the size changes revealed by the fossil studies are all caused solely by proximate growth and survivorship differences among sites.

Some correlative cases would seem instructive. The islands in the vicinity of Madagascar have been studied by Arnold and associates [14, 365, 15, 16, 48].

Aldabra had crocodiles, a large iguanid, two or three geckos, and two skinks that no longer occur there. Yet Arnold points out that human influence is not likely to be involved in these extinctions as there is no evidence of permanent settlement, and close relatives of these extinct species coexist with man elsewhere. Arnold attributes these extinctions to a drastic reduction in the size of the atoll following the sea-level rise at the end of the Pleistocene. In contrast, early colonists to Mauritius almost surely hunted the large turtles there to extinction, and a number of smaller reptiles that were known on Mauritius are now restricted to tiny shelf islands nearby, presumably as a result of human activities. The key point is that the effect of man depends on both the island and the man.

Fewer extinctions of lizards are known in the Lesser Antilles than on the islands near Madagascar. As discussed earlier, only the *Leiocephalus* line has become extinct, one species from the Antigua Bank, and one from the Anguilla Bank. These extinctions appear to predate the arrival even of lithic man and, like Aldabra, seem attributable to the last sea-level change. The only other known extinction of a small lizard in the Lesser Antilles is *A. pogus*, a member of the wattsi series on Anguilla, as discussed extensively above. Human activities would seem deleterious to *A. pogus*, and those activities, combined with the strong competition it received from *A. gingivinus*, have evidently caused the extinction. But the wattsi-series populations on the other banks, where presumably an equal or greater degree of human disturbance has taken place, have not become extinct. The difference is that the wattsi form on Anguilla was receiving strong competition from another anole, while the wattsi-like forms on the other island banks are not.

Similarly, Richman et al. [273] concluded on the basis of an extensive review, combined with a detailed analysis of the lizards on the Baja California Islands and the South Australian Islands, that human disturbance primarily exacerbates the processes of extinction that are already in place. The most obvious effect of man on anoles is to introduce them to new locations. The great number of known introductions of anoles contrasts with the scarcity of extinctions of anoles.

2.5.4 Phylogenetic analysis

Comparisons among species, islands, and sites often have poor statistical power because the extent of shared phylogenetic influences are not quantified. Varying degrees of relatedness among taxa imply that the data points are not fully independent, and make assessing the "degrees of freedom" for a statistical test difficult. During the last five years there have been improvements in how comparative studies are done in ecology and evolutionary biology [150, 151, 243, 133, 221]. The starting point for an analysis of ecological differences among species that takes into account their interrelatedness is a diagram called a "cladogram," a branching diagram that indicates the hypothesized relative timing of when the various lineages among the species split apart from one another. Chapter 3 details a cladogram for the eastern Caribbean anoles that will be discussed then in relation to reconstructing the geologic origin of the eastern Caribbean.

Recently, this cladogram has been used by Losos [193] to see if there is phylogenetic evidence of character displacement. Losos [193, p. 558] writes of the body-size pattern in the eastern Caribbean that "two processes could produce such a pattern: size adjustment (or character displacement), in which similar-sized species evolve in different directions in sympatry; and size-assortment, in which only different-sized species can successfully colonize the same island together." Losos estimated the expected body size of hypothetical ancestors using a minimum evolution algorithm based on the criterion of squared change parsimony [151]. Losos found, for the northern Lesser Antilles, that character displacement might have occurred only once and that the relative rarity of character displacement suggested that size assortment is responsible for the biogeographic body-size patterns. In the same vein, Losos has written concerning the northern Lesser Antilles [195]: all the small species are members of a monophyletic clade and, similarly, the large species on the two-species islands form a second monophyletic group . . . Consequently, the most parsimonious interpretation is that large and small size evolved only once in these clades . . . Based on this reasoning, the transition from one- to two-species islands occurred simultaneously with the evolution of large and small size — exactly the prediction of the character-displacement hypothesis. However, the data only provide evidence for one instance of character displacement, which suggests that the widespread pattern of size dissimilarity has resulted from size assortment subsequent to the evolution of size differences.

Although Losos [194] has written as a proponent of the character displacement hypothesis, it is worth emphasizing that Losos' own defense of character displacement does not support, and is much weaker than, the original proposal for character displacement by Williams [383] in which *each* two-species island, or at least each two-species island bank, was hypothesized independently to develop size differences by character displacement. All Losos is claiming is that character displacement has occurred once, and thereafter invasions by the already differentiated species onto new islands established communities that possessed size differences in them from the beginning.

In fact, Losos' phylogenetic analysis does not resolve whether the differentiation of the small- and large-sized clades was actually caused by the divergent coevolution of syntopic species (species living together). A two-species island could be produced by a propagule from a mesic habitat on one solitary island, together with a propagule from a xeric habitat on another solitary island. The phylogenetic analysis would appear as character displacement on the two-species island, even though the propagules would not have evolved much of their size differences in response to competition from one another but, instead, would have evolved most of their size differences as independent responses to the habitats from which they originated.

In a follow-up phylogenetic analysis by Miles and Dunham [222], the size characters were coded using "generalized gap coding" [12] as appropriate for quantitative characters, and evidence of large and intermediate species becoming small with increasing phylogenetic distance from the root node was found. Also, a reexamination of the cladogram showed two instances where intermediate-sized descendents evolved from large-sized ancestors.

2.6 Hypotheses evaluated

Now let's integrate the historical evidence into the other evidence we have accumulated and come to some conclusions about how the biogeographic patterns formed and about how communities have been assembled in the Lesser Antilles.

The pure invasion scenario postulates that a two-species island is formed from the invasion of opposite extremes from the range of observed solitary sizes. The body-size differences on a two-species island is thus its initial (or primitive) state. These differences are acquired in allopatry and permit an invasion because they exceed the Hutchinsonian ratio of 1.3. But because two-species islands have species both smaller than the small extreme solitary size, and larger than the large extreme solitary size, further evolution has evidently taken place that this scenario cannot explain. I think this scenario is best at accounting for how the first two-species island formed, but thereafter one of the other scenarios seems better.

The character displacement scenario postulates that a two-species island is formed from the invasion of two solitary-sized species, each of which diverges from the other. The body-size difference on a two-species island is thus a final (or derived) state. The differences are acquired *and* maintained in sympatry. The main problem with this hypothesis is that the competition that drives each species to diverge from each other also prevents the invasion of a second solitary-sized species onto an island that already has a solitary-sized resident. The data on invasions show that failure is more likely if a resident has a similar size to the invader. Data on size changes from fossils also consistently show large species becoming smaller, not medium-sized species becoming larger. Phylogenetic evidence shows at most one instance of character displacement in the northern Lesser Antilles. Still, these difficulties notwithstanding, character displacement could enhance a body-size difference already large enough to have permitted a successful invasion.

The parallel evolution scenario postulates that a two-species island is initially formed by the invasion of a very big or very small species, and then both species evolve in parallel, either down or up. For example, if the wattsi series is considered as the original residents in the northern Lesser Antilles, and the bimaculatus series the invaders, then the size changes in these anoles appear consistent with a parallel evolution scenario: All are evolving down in size. The problem is that there is no source for the first very large species. The large species on a typical two-species island is larger than the largest solitary anole (other than Marie Galante's), and there's no place on a present-day solitary island this first invader could have come from and still be as large as the larger species on a two-species island. Nonetheless, this scenario might be fine when applied to the second two-species island. If a solitary anole is invaded by the larger species from an already-existing two-species island, both species on the new two-species island might indeed evolve downwards in parallel. Or a variant of this scenario is that the large species could stay as it is while the solitary anole shifts downwards all by itself.

Indeed, a synthetic scenario is possible that combines the best parts of all of these: The first two-species island is formed by invasion of opposite extremes from the range of solitary sizes. The body-size differences on this island are enhanced by character displacement. Then the large species from this island crosses to more solitary islands, where downwards evolution takes place by the original resident. A qualification is that the present-day sizes may underestimate the size range previously available across habitats on the solitary islands. If the largest solitary anoles have been getting smaller, as have the largest anoles on the two-species islands, then the previous size range of solitary anoles may be greater than the present-day range of 1.3. If so, less character displacement would be needed to attain the present-day ratio of 1.7 on the two-species islands.

But what about the exceptional islands of Marie Galante and St. Martin? The taxon loop, our remaining scenario, does the best job of explaining the exceptional islands. If the competition is asymmetric and the width of the carrying capacity function narrow, then a larger invader may overtake the original resident, causing its extinction. St. Martin would seem to be precisely this situation—an invading *A. gingivinus* causing the extinction of the resident *A. pogus*, as already has happened on Anguilla. In this scenario Marie Galante would be the stage following the extinction of a smaller species, and *A. ferreus* is viewed as on its way down to the solitary size. The two problems with this scenario are, first, that it is fragile theoretically and, second, that the body sizes on Marie Galante and St. Martin don't work out right. If Marie Galante and St. Martin were ecologically identical islands, and if Marie Galante represents a more advanced stage of a taxon loop than St. Martin, then *A. ferreus* should be smaller than *A. gingivinus*. Yet the reverse is true. Unfortunately though, if the taxon loop is rejected we are left with no explanation for the exceptional islands that is not ad hoc. For example, some habitat change may have happened on these islands but not the others that disproportionately affects the smaller species. Also, *A. ferreus* is possibly the oldest solitary species in the eastern Caribbean and, as such, is the only species to have had time to expand its niche width to encompass the niche space normally occupied by two species. So, to explain the exceptions, one must entertain either special-case explanations for each of these islands or entertain a taxon loop scenario that has theoretical and quantitative problems of its own.

Thus, substantial data and theory can account for the main biogeographic pattern of *Anolis* body sizes in the northeastern Caribbean but the status of the exceptional islands still remains unsettled.

Concerning particular colonization events, the northern Lesser Antilles appears to have broken off from the Guadeloupe/Dominica region carrying the ancestor of the wattsi lineage. This ancestral population then broke up into the three wattsi-series species of *A. pogus*, *A. schwartzi*, and *A. wattsi* as the three distinct banks of the northern Lesser Antilles formed during the eastward motion of the Caribbean Plate. The Antigua, St. Kitts, and Anguilla Banks were then colonized through overwater dispersal by a montane form of *A. marmoratus* from Guadeloupe in accord with the prevailing ocean currents. These could be separate colonization events or a stepping-stone sequence with Antigua first, followed by St. Kitts. Also, humans may have aided the introduction of *A. gingivinus* to the Anguilla Bank from Saba. In the southern Lesser Antilles, an attractive scenario has a montane form of *A. trinitatus* from St. Vincent colonizing Grenada to become *A. richardi*, which in turn recolonizes back to St. Vincent to become *A. griseus*. Of course, specifics such as these must be regarded as completely conjectural until more fossil and phylogenetic analysis has taken place.

The Greater Antilles may, or may not, be viewed as extensions of the Lesser Antilles, depending on the significance of the exceptions to the size rules. If many two-species islands drop down into a single-species state, then small islands are locked in a low-diversity regime. Alternatively, if the two-species state is relatively permanent, then it is available for buildup to three species and higher. Even so, the addition of each succeeding species takes longer: Even a successful invader is inhibited to some degree by diffuse competition from the existing residents, and also the invasion process is increasingly selective implying an increasingly low probability that a suitably sized invader exists anywhere in the biogeographic region. My reading of the matter is that buildup much beyond two species by sequential addition is not realistic. I think buildup to a large community is brought about by blockwise addition, as plate tectonic units come in contact allowing entire communities the possibility of cross-invasion, a theme pursued further in the next chapter. Thus, it may not be correct to view island faunas as precursors to large continental faunas or, for that matter, as miniature or scale models of continents. The relation between an island fauna and a continental fauna may be analogous to that between a tissue culture and an organism. A tissue culture is extracted from an organism, grown in isolation, and used to spotlight a particular process or mechanism. Similarly, an island's fauna may be an extraction from a complex fauna that develops in isolation and that exaggerates certain processes present, but perhaps not as influential, in complex continental faunas.

The scenarios we have entertained to explain community assembly in the eastern Caribbean are important for conservation biology. Inherent in the idea of character displacement hypothesis is that species coevolve mutual accommodation with one another in a local site, such as an island. Suppose a conservation organization buys an island as a reserve. One could imagine stocking it with various species of anoles (and other organisms) and returning much later, say, 10,000 years, to find that all the species had coevolved to get along better with one another, as in avoiding competition by partitioning resources. Moreover, functional diversity has increased through time. Initially we have some similar species that could probably be substituted for one another, and after character displacement the species have differentiated from one another ecologically. The diversity of niches, the variety of "occupations" for species, has increased as an endogenous process.

Instead, we have come to view diversity as a property of the region—the entire system of islands, not the property of one island. The reason why biodiversity develops in a particular spot is because that spot happened to be near other places where different niches were favored. Then when dispersal allows the species from different places to mix with one another, compatible subsets emerged that are more diverse than any island could produce on its own. To conserve this system, it's not necessary or even desirable to purchase one island and lock it up from intrusion for 10,000 years. Instead, a system of habitats among islands is needed.

The parallel evolution and taxon loop hypotheses differ in whether extinction is provoked by external intervention or happens as a natural biological process. By the parallel evolution hypothesis, either geological or human habitat modification sets in motion the processes leading ultimately to an extinction. Protecting diversity here requires great attention to ameliorating habitat modification. By the taxon loop hypothesis, extinctions occur naturally in the absence of environmental change, and, to conserve diversity, replacement species must be actively introduced.

The realization that local biodiversity reflects the habitat diversity of the biogeographic region brings us to the next scale, the landscape scale.

2.7 Biogeography of habitat use

Are the islands of the eastern Caribbean ecological replicates of one another, are they homogeneous with respect to process and outcome? For food resources, regularities were discovered across all the islands, as reflected to body size. In contrast, for space resources, the islands fall into two distinct classes that coincide with the phylogenetic distinction between the bimaculatus and roquet communities of the northern and southern Lesser Antilles. Also, the use of space on a landscape scale is qualitatively different between the bimaculatus and roquet communities.

2.7.1 Niche axis complementarity

The axes by which anoles in the Lesser Antilles partition space are height and microclimate. On bimaculatus two-species islands, the anole species perch in different heights: The smaller species perches nearer the ground than the larger species. Table 2.20 shows the average perch height for the various species. In the bimaculatus communities the average perch height is less than 1 meter for the smaller species, while the larger species is higher than 1 meter or more. While these values can depend on the height of the available vegetation, the separation

Table 2.20 Niche positions on space-use axes.

Island	Species	Perch height (SE)	Perch GBTI (SE)
	Bimaculatus communities		
St. Kitts	*A. schwartzi*	0.12 (0.06)	34.0 (1.3)
	A. bimaculatus	0.91 (0.09)	32.5 (0.7)
Barbuda	*A. forresti*	0.61 (0.06)	33.7 (0.6)
	A. leachi	1.77 (0.06)	34.1 (0.8)
St. Martin	*A. pogus*	0.79 (0.03)	28.2 (0.5)
	A. gingivinus	1.46 (0.06)	29.4 (1.2)
St. Eustatius	*A. schwartzi*	0.03 (0.03)	30.4 (0.2)
	A. bimaculatus	1.52 (0.12)	30.1 (0.2)
	Roquet communities		
Grenada	*A. aeneus*	1.37 (0.06)	30.1 (0.4)
	A. richardi	1.37 (0.06)	28.3 (0.3)
St. Vincent	*A. trinitatis*	1.13 (0.06)	31.0 (0.8)
	A. griseus	1.25 (0.06)	29.0 (0.5)

in perch height is usually about 1 to 1.5 meters, a difference that is conspicuous in the field. Roquet two-species communities show no consistent difference in perch height across sites, although at a given site one species may perch consistently above the other, depending on the height at which the sunny and shady places are principally found.

The other axis used for partitioning space pertains to the microclimate. In an early study of resource partitioning, Schoener and Gorman [312] observed that the species on Grenada usually had different body temperatures because one species usually perches in sunny spots while the other perches in shady spots. The temperature of the lizard represents the net result of both abiotic and biotic paths of heat exchange. To quantify the abiotic component, the temperature of an inanimate lizard-sized reference object can be used. Its temperature reflects the balance of incident solar radiation and convective heat loss (or gain) from the object to the surrounding air. By measuring air temperature, wind speed, and solar radiation to a perch position, the temperature of a reference object can be calculated [293]. Alternatively, a physical object can actually be constructed and placed at the spot where a lizard perches and its temperature determined. Either way, the implications for body temperature of the overall microclimate at a perch position can be measured. Table 2.20 presents an index, called the "Grey Body Temperature Index" (GBTI) for the spots at which lizards perched between 9:30 am and 3:00 pm during the day for July and August. The units are °C. In the roquet communities, the smaller species consistently perches at hotter spots than the larger species, the difference being about 2°C. In contrast, the spots used by both species in the bimaculatus communities have the same GBTI, and also their body temperatures are not significantly different.

Thus on all the two-species islands, the two species tend to perch in different types of places. In the bimaculatus communities perch height is the axis along which the separation occurs, while in the roquet communities perch temperature is the axis along which the separation occurs. This phenomenon is called "niche axis complementarity."

Although the bimaculatus anoles do not differ in the microclimates they actually use, as mentioned earlier, the two species on an island do generally differ in their tolerance to direct sunlight for long periods. When a lizard is held in the open sun for a period of time, say five minutes, it gradually heats up and at some temperature begins to pant, a sign of heat stress. The smaller species of the bimaculatus two-species islands generally are less tolerant to direct sun, and begin panting at a lower body temperature, than the larger species. This fact presumably limits the activity time of the smaller species in xeric conditions to the cool times of day. Notice that this is the reverse of the roquet situation where the smaller species is the one perching in the hotter microsites.

The anoles of the bimaculatus and roquet groups also seem to differ in shape. Both anoles of St. Kitts, for example, have a hind leg that is about 55% of the snout-vent length. On Grenada, the larger species has a hind leg that is about 60% the snout-vent length, while the smaller species has a hind leg about 50% the snout-vent length. This fact has been interpreted to indicate that the two species on Grenada use the habitat differently, while both anoles of St. Kitts use the habitat in the same way [226].

No comparable data have been obtained for all the one-species islands, although thermal and water-loss studies have been done on Dominica, Guadeloupe, and Martinique [42, 153, 146, 145]. Also, the distribution and abundance of the solitary anole across habitats on Dominica has been measured [49]. It is not known whether all the regular solitary-sized anoles are also the same in their use of space. *A. acutus* of St. Croix appears to me much more arboreal than *A. lividus* of Montserrat, for example. There may very well be heterogeneity in the use of space among the solitary anoles, just as there is among the two-species communities.

Traditionally in lizard (and vertebrate) ecology, both space and food are thought of as distinct resources, both of which may be partitioned independently, with prey size (as reflected in body size) the basis for partitioning food, and perch height or microclimate the basis for partitioning space. This interpretation may have been superseded, however, by the experiments mentioned earlier on St. Eustatius, in which competition did not increase when *A. bimaculatus* was experimentally forced to perch closer than normal to *A. schwartzi*. Thus, space may not be a limiting resource, and perching in different places may not represent a niche partitioning of space in the sense of competition theory.

2.7.2 Within- and between-habitat diversity

The landscape scale is the scale that is perceived as from an airplane or in a satellite image. This scale tends to reveal spatial patches of habitats or local communities. A given expanse of physical environment might be covered uniformly by one community type or by different community types, each associated with specific topographic features such as slope, orientation, and angle to the sun. Similarly, the total number of species from a genus may be distributed throughout a region in ways that represent mixtures of two extreme patterns. In one extreme, all the species have completely overlapping ranges; all the species co-occur everywhere in the region. At the other extreme, all the species have nonoverlapping ranges; each species has a piece of the region to itself and there is no coexistence anywhere. Between

Table 2.21 Abundance of *Anolis* lizards in roquet communities.

Grenada	A. aeneus	A. richardi
Les Advocate (forest)	1.3 (0)	16.3 (2.5)
Beach Forest (forest)	10.8 (2.6)	18.0 (2.6)
Junction (woods)	4.1 (0.9) adult	0
	29.0 (8.6) hatch	
Coral Slope (scrub)	Rare	0

St. Vincent	A. trinitatis	A. griseus
Kings Hill (forest)	5 (3)	55 (17)
Layou (woods)	50 (10)	Rare

these extremes one can imagine a spectrum of possibilities. The "between-habitat diversity" of a region refers to the number of species in the region that can be identified with different places, while the "within-habitat diversity" refers to the number of species that co-occur in the same habitat[17] [380, 276]. Faunas are known to differ in how the regional species diversity is apportioned into the within- and between-habitat components. Cody [66] has shown for large regions, such as deserts and Mediterranean-shrub/savannas, that the diversity of birds is apportioned differently into the within- and between habitat components on different continents, as has Pianka [247] for desert lizard communities.

The two-species communities in the Lesser Antilles also appear to differ fundamentally in how the diversity is apportioned into within- and between-habitat components. This point is difficult to document without a very large expeditionary effort that can yield a census of the lizards at a regular spacing of grid points across an entire island. Tables 2.21–22 document the situation based on a more modest effort [289]. In the tables, "rare" means sighted adjacent to quadrat, or only one marked in quadrat; "present" means two or more marked inside quadrat. The standard error of estimate is given in parentheses, and the units are number of lizards per 100 m². Censuses were conducted during August 18 to 28, 1976, on Barbuda, September 2 to 7, 1976, on Grenada, August 23 to September 9, 1977, on St. Martin, August 5 to 20, 1978, on St. Kitts, July 26 to 30, 1979, on St. Eustatius, and supplemented with unpublished data from G. Gorman and J. Lynch for St. Vincent during July 17 to 29, 1975.

Table 2.21 shows the abundance of anoles at several sites in Grenada and St. Vincent, both roquet islands. In forest habitat the larger anole, *A. richardi* of Grenada and *A. griseus* of St. Vincent, is numerically more abundant than the smaller species. But at xeric and open sites near sea level, the numerical dominance

---

[17]Other terms are close to these in meaning. Species whose ranges overlap are said to be "sympatric" in the area where the ranges intersect, and "allopatric" where the ranges do not overlap. Within the area of sympatry, species are said to be "syntopic" if the individuals of the different species encounter each other regularly, and are "allotopic" if each uses sufficiently different microhabitats that the individuals have little potential for direct interaction. Also, the within-habitat component of diversity is sometimes called "alpha diversity" and the between-habitat component called "beta diversity."

Table 2.22 Abundance of *Anolis* lizards in bimaculatus communities.

Barbuda	A. forresti	A. leachi
All-in-Well (woods)	35.1 (3.5)	5.9 (1.1)
Junction (woods)	18.8 (1.4)	3.1 (0.6)
Cross Path (scrub)	4.6 (0.8)	1.2 (0.2)
Coral Slope (scrub)	2.7 (0.7)	0

St. Kitts	A. schwartzi	A. bimaculatus
Top Antenna Road (woods)	49.7 (4.5)	Rare
Med. Antenna Road (forest)	29.2 (4)	Present
West Farm Crest (forest)	76.3 (8.8)	Rare
Monkey Site (scrub)	29.1 (1.9)	Present
Little Hill (scrub)	Present	Present

St. Eustatius	A. schwartzi	A. bimaculatus
Quill Top (woods)	25.3 (1.4) adult 39.3 (11.9) hatch	Rare
Central Hill (woods)	32.7 (2.4)	21.3 (5.9)
Cannon Point (scrub)	0	11.9 (2.0)

St. Martin	A. pogus	A. gingivinus
Pic du Paradis (woods)	10.5 (0.7)	8.4 (2.6)
Medium Pic (woods)	17.7 (0.7)	10.2 (2)
St. Peter Mt (woods)	16.9 (0.8)	19.7 (1.4)
Boundary (scrub)	0	50.1 (2.2)
Naked Boy Mt (scrub)	0	43.2 (3.4)
Well (woods)	0	42.4 (1.9)
Point Blanc (scrub)	0	6.4 (0.5)

reverses, with the smaller species being more abundant. At sufficiently xeric sites the larger species is apparently absent altogether. The somewhat idealized picture to emerge is that on the roquet islands the larger species is numerically dominant throughout a more or less circular region of forest at the island's center, while the smaller species is numerically dominant in a strip along the circumference of the island. The range of both species broadly overlap, both species co-occur at most sites, and the smaller species is probably present everywhere on the island, even though numerically in the minority for forest habitats. Thus, the roquet two-islands exhibit some of both the within- and between-habitat components of diversity.

Table 2.22 shows the abundance of anoles at sites on four of the two-species bimaculatus communities. Two patterns seem evident. Recall that Barbuda and St. Kitts have species with large differences in body size. On these islands both species are present together in almost all habitats throughout the islands. The smaller species is more abundant than the larger species everywhere. If a site is xeric enough to have only low bushes then the total abundance of anoles is so low that the larger species may be absent, as in the Coral Slope site on Barbuda, even though individuals of the larger species may be found, say, 500 m away. On St. Kitts, however, I have not located a site where both species did not occur, even though the abundance of each might be too low to census quantitatively. Thus, on these islands there is essentially no between-habitat component of diversity. All the habitats on these islands have the same diversity of anoles.

Table 2.22 also shows the abundance at sites on St. Eustatius and St. Martin. Recall that these islands have two species with relatively little difference in body size, especially so for St. Martin. On these islands the smaller species mostly populates hills in the center of the island, and, again, especially so on St. Martin. In the central hills, the smaller species is numerically more abundant than the larger species, as on the other bimaculatus islands. But the smaller species drops out toward the xeric habitat near sea level. This pattern is very conspicuous on St. Martin where *A. pogus* is restricted to the central hills. Thus, these islands do have some degree of between-habitat diversity, but in a pattern opposite to the roquet islands in that the smaller species is numerically dominant in forest habitat, unlike the roquet islands where the larger species is numerically dominant in forest habitat.

Thus, the bimaculatus communities tend *not* to have any between-habitat diversity, in contrast to the roquet islands, if the body size difference in the co-occurring anoles is large, and to have a between-habitat pattern of diversity that is qualitatively opposite to roquet situation if the body size difference in the co-occurring anoles is relatively slight.

2.8 Theory of habitat use

2.8.1 Multiple-niche axes

A multiple-axis version of niche theory has been investigated using multivariate bell-shaped K curves and α functions [237]. The key results for multiple axes and two species are: First, the only coevolutionary equilibrium niche positions lie *on* the axes. For example, if two axes are available for resource partitioning, only one axis is actually used in any coevolutionary equilibrium. Second, all of these equilibria, or some subset of them, each corresponding to a different axis, may be simultaneously stable in a coevolutionary sense. Only one equilibrium is stable when substantially more separation is possible along this axes than along any other. However, if the separation achieved on another axis is about the same, then both equilibria are simultaneously stable. In this case, the axis that is involved in the resource partitioning observed in a particular system depends on the initial condition for that system.

This result may be relevant to the niche complementarity of Lesser Antillean anoles. First, perch height and GBTI do seem to be independent axes; that is, high perches are not necessarily hotter than low perches, or vice versa, although high perches generally have a higher illumination, but also steady breezes. The higher wind speed at high perches compensates for the higher illumination there yielding approximately the same distribution of GBTI as at lower perches [293]. Second, neither perch height nor GBTI are necessarily niche axes at all, because space may not be a limiting resource. But, *if* they are valid axes, then the niche complementarity shown by the bimaculatus and roquet two-species communities may represent each of these groups attaining different alternative and simultaneously stable coevolutionary equilibria. According to this hypothesis, the presence of partitioning with respect to one or another axis on an island is an accident of the initial dispositions of the colonizing lineages.

Why would the bimaculatus group be disposed to partitioning in the perch height dimension and the roquet group in the microclimate dimension? Possibly a predisposition toward microclimate partitioning in the roquet group is caused by the prior evolution of the habitat segregation that exists there. Given the habitat segregation (one species numerically predominating in open habitats, the other numerically predominating in forests), then the physiological differences that underly these habitat differences can predispose the use of microclimate for the local resource partitioning too.

In contrast, the colonists in the bimaculatus-group islands may have originally employed perch height as one of the major axes to partition space, as currently observed in the Puerto Rican anole fauna. These colonists may have brought this disposition with them.

As mentioned earlier though, the experimental competition studies suggest that only food, and not space, is a limiting resource. If so, the niche axis complementarity between perch positions in the roquet and bimaculatus two-species islands has no special coevolutionary significance, but could be useful as a taxonomic character.

2.8.2 Habitat segregation

Density-dependent selection theory has been extended to a spatially varying environment in a first step toward understanding the evolution of habitat segregation [289, 140]. The model is spatially explicit, involves partial differential equations with both space and time as the independent variables, allele frequencies and abundance as the dependent variables, and includes dispersal. The population is assumed initially to be fixed for an allele, A, whose carrying capacity is uniform throughout some geographical region. Then an allele, a, arises that has a higher carrying capacity in one subregion at the expense of a lower carrying capacity in the remainder of the region. If a enters the population, then the population will depart from its initial pattern of uniform abundance throughout the region and instead become concentrated in the subregion where a is advantageous, while becoming comparatively rare in the remaining subregion. Thus, the entrance of a into the population marks the first conceptual step in the evolution of habitat segregation. If two species evolve in this way, but into opposite habitats, then habitat segregation results.

The intersting result of this theory is that habitat specialization evolves depending on the geometry of the region, particularly the size of the subregion within which a confers advantage. If this region is large enough, then a enters the population regardless of how bad a is in the remainder of the region. If the region is somewhat smaller, but not too small, then a enters the population only if it is good enough in the favored subregion relative to how bad it is in the remainder of the region. Finally, if the size of the subregion where a confers an advantage is small enough, then no matter how good it is there, a simply won't enter the population. So, as a broad generality, the evolution of habitat segregation is favored if both subregions are more or less equal in area. This would allow one species to adapt to the habitat of one subregion and the other species to the habitat of the other subregion. In contrast, if one habitat type has much more area than the other, then both species will adapt to the more common habitat type while abandoning the rare habitat type.

The two-species *Anolis* communities of the bimaculatus-group islands don't exhibit habitat segregation, whereas those in the roquet group do. In the roquet group, one species predominates in shade forest, and the other predominates in open woods and scrub. Why? For some of the bimaculatus-group islands there is trivially little variation in habitat to begin with. Barbuda, Antigua, St. Martin, and St. Eustatius all are quite flat and arid, with little other than open woods and scrub. But St. Kitts in the bimaculatus-group islands does possess a mountain as tall as those in the roquet-group islands of St. Vincent and Grenada.

The principal geometric difference between St. Kitts and the roquet islands of St. Vincent and Grenada is the fraction of the island's area that is mountainous. The mountains of St. Kitts form a narrow band in the center of the island and are absent altogether from its southern panhandle, whereas mountainous forest comprises about half the area of both St. Vincent and Grenada. Assuming St. Kitts forest is comparable to Grenada forest and St. Kitts lowland to Grenada lowland, then the issue becomes one of relative areas. Possibly on St. Kitts, and more so for the remaining bimaculatus-group islands, the area of deep-shade forest is too small to permit the evolution of a specialist to this habitat type, implying that both species in the bimaculatus communities are primarily adapted to arid lowland habitat. In contrast, the relatively large area of deep-shade forest in Grenada and St. Vincent may have permitted the evolution of a specialist to this habitat type. Presumably, the evolution of habitat specialization in one of the two competing species then facilitates the reciprocal evolution of specialization to the complementary habitat type by the other species.

2.9 Discussion

Islands serve as laboratories where the workings of natural ecosystems can be discovered, as first illustrated by Darwin's insights about evolution from the fauna of the Galapagos Islands. The West Indies are especially interesting because they are more numerous and older than the well-studied Galapagos and Hawaiian Islands. Also, they feature lizards rather than birds as the main insectivorous consumers in their terrestrial communities. Findings from research on islands, including the West Indies, are relevant to conservation biology and potentially to the economic analysis of biodiversity and ecosystem function.

What have we learned from the anoles of the West Indies, particularly the

Lesser Antilles? Anoles are limited primarily by food. Species of anoles must be different in body length by a factor of about 1.3 to coexist, because different body sizes consume different prey sizes and thereby partition the resouces and avoid competition. Hence, a structure is evident in ecological communities of *Anolis* lizards: The body length of each species can be identified with an occupation, or niche, that differs from the occupations of the other species it lives with. The total resource use by the *Anolis* community is partitioned among these niches. This kind of structure also occurs in other animal communities from bumblebees, desert rodents, and lake fish, to a great many bird communities. Yet, this structure is not found in most marine communities where species often coexist that are functionally equivalent in their resource use, although perhaps different in other ways including life cycle and sensitivity to disturbance, predation, and grazing. Nonetheless, a community structure based on partitioning food resources with respect to prey size is particularly clear in the *Anolis* lizards of the eastern Caribbean.

The differences in body size that enable anole species to coexist mostly orginate as adaptations to ancestral habitats, not as adaptations to each other. When two solitary species of anoles differ in body size because each comes from a different habitat, say, one from a mesic and the other from a xeric habitat, then these body size differences allow each to invade the other's range when the physical opportunity arises. Thereafter these differences explain how the species can coexist, even though little if any of the differences arose through in situ coevolution to reduce competition. Hence, local species diversity reflects regional habitat diversity, and to conserve biodiversity, one must preserve the region's diversity of habitats. Finally, the anoles of the northern and southern Lesser Antilles show different patterns of landscape organization. In the north, both species on the two-species islands vary together in abundance: Both are more abundant in wet than dry habitat and have about the same relative abundance everywhere. In the south, the species show complementary habitat use, with one species being more abundant than the other at sea level, and the reverse in the forests of the central hills.

3

Origin of the Caribbean

The *Anolis* lizards found throughout the islands of the Caribbean and in Central America may serve as "living strata." Here, information about eastern Caribbean anoles is offered to help reconstruct the origin of the Lesser Antillean Island Arc in relation to Puerto Rico and the Netherlands-Venezuelan Antilles. This chapter is taken primarily from [285]; a preliminary account appeared in [288].

Today the Caribbean region is a tectonic plate of its own, distinct from the North and South American Plates, and it is moving east relative to the Americas (cf. [85, 55, 212]). The leading edge of this plate is in the Atlantic beyond the the Lesser Antilles, where the North and South American Plates dip down into the mantle underneath the Caribbean Plate forming a Benioff zone. The trailing edge of the plate lies in the Pacific Ocean offshore of Central America from Hondurus to Panama. The northern margin of the Caribbean Plate extends from the Puerto Rico trench north of Puerto Rico, through Hispaniola, the Cayman Trough, and though Honduras. The motion along the northern margin is left-lateral.[1] The southern margin extends from Trinidad, along the coastal margin of Venezuela, and intersects a complex of faults west of Lake Maracaibo in a region influenced by the collision of Panama with Columbia. The motion along the southern margin is right-lateral. Finally, the sea floor of the Caribbean is coated by a thick layer of basalt, a so-called "flood basalt" that is also often called the "B" layer.

The Caribbean Plate poses many questions, most fundamentally, why it exists at all. The intuitive appeal of plate tectonics is the idea of rigid pieces of a jigsaw puzzle that can be moved around to synthesize the continents of the past. The fit of the Americas with Europe and Africa is the perfect example. Between these continents today lies the great mid-Atlantic spreading ridge. But alas, if one moves North and South America together, they don't match up very well. A lot of unexplained material in the Caribbean and Central America has come from somewhere, or somehow been made. Moreover, no spreading ridge exists today running east-west along the middle of the Caribbean analogous to the spreading ridge running north-south down the middle of the Atlantic. Instead, between North and South America lies a tectonic plate with a life of its own, and the basic question is why this plate is there at all, and where it came from.

Of course, other more detailed issues arise as well. Are some of the islands older than others, and if so which are the oldest? Have the islands always been where they are now, and if not where were they? Have some of the islands been

---

[1]Left-lateral means that if one stands next to a fault and looks to the other side after an earthquake, the other side has moved to the left; and conversely for right-lateral slip.

submerged, and if so, how much and what parts? Finally, what sort of habitat has been available for organisms in the Caribbean through geological time?

The approach taken in this chapter is to use the organisms themselves as if they were strata. The idea is simple. We know that when one population is somehow split into two subpopulations, then each subpopulation goes its own evolutionary way. If one island, for example, splits into two distinct islands, then the population on each fragment evolves its own characteristics independently of the other. And if one of these subpopulations again splits, then these too will go their own evolutionary ways. A process of successive fragmentation like this will generate a treelike diagram of how all the populations are related to one another. Now let's run this in reverse. If we can somehow determine the treelike diagram of relationships among a set of populations, then the sequence, and perhaps even the timing, of when they broke apart from one another can be reconstructed. If geologic processes such as plate-tectonic motion are responsible for the breaking apart of the populations, then the biological lineage relationships become relevant to reconstructing the plate-tectonic processes involved.

Two reasons make using lineage relationships to reconstruct plate tectonic motion in the Caribbean attractive. The first is that the technology for determining lineage relationships has vastly improved during the last 20 years. When I was a student[2] lineage relationships among species were not to be taken too seriously. Usually some authority pronounced who was related to whom, and that was that. Today, lineage diagrams are based on the analysis of large data sets and are objective and repeatable. The lineage diagram is understood to be an *estimate* of the relationship that may be correct as is or that may change as more data are obtained. An analytical approach called "cladistics" is increasingly used to construct a lineage diagram, and the resulting diagram is called a "cladogram." Biochemical characteristics are increasingly popular as supplements to morphological traits. There aren't enough morphological traits to go around, and biochemical traits greatly add statistical power. The second reason for using lineage to aid in reconstructing Caribbean tectonics is that, as we shall now see, *Anolis* seems perfectly suited to this purpose.

For a biological population to mark a geologic entity, such as an island or island bank, it must be a ready colonist: It must rebuff subsequent colonizations from adjacent geological entities, for otherwise the labels would mix, and it must not be prone to extinction. Anoles are hardy survivors on tropical cays; ecological competition between anoles for limiting resources retards cross-invasion between adjacent islands and prevents cross-invasion altogether between species of the same body size: And the huge population size on a Lesser Antillean island (about 10^8 for a 400 km^2 island, assuming 0.25 lizards per m^2) precludes chance extinction and increases the likelihood of contributing a propagule to newly opened habitat. Also, anoles are diverse and thereby provide many labels. There are over 300 species of *Anolis*, about half of which are in the West Indies and half in Central and northern South America.

Moreover, anoles have been in the Caribbean theater for a long time. A complete fossil *Anolis* encased in early Miocene amber (ca. 20 to 23 Ma[3]) has been found in the Cordillera Septentrional of the Dominican Republic [277]. It may be even older than originally reported, lower Oligocene or upper Eocene (35 to 40 Ma) [252]. The specimen is a juvenile lizard indistinguishable from the green

---

[2]I know this dates me.

[3]Ma means 1 million years before present.

anoles now living in Hispaniola, *A. chlorocyanus* and *A. aliniger*. Skin fragments from another fossil anole have been described from Chiapas, Mexico; its amber has been dated as late Oligocene to early Miocene [179]. Finally, the lineage (clade) consisting of the present-day Iguanidae, Agamidae, and Chamaeleonidae, and including *Anolis*, is an old line, extending back at least into the middle Jurassic (175 Ma) [106]. In contrast, old rock strata in the Caribbean are scarce, as Jurassic and Cretaceous materials have been lost to subduction or covered by more recent magmatic activity. Thus, anoles have been in place as the Caribbean itself has formed, and the systematics, ecology, and biogeography of these lizards may therefore provide sorely needed clues about the origin of the Caribbean to supplement geologic data. To present this information, the the biological evidence is first developed, and then integrated with geological data.

For the eastern Caribbean the key conclusions offered are:

1. The northeastern and southeastern Caribbean have had separate geologic origins. The line separating these two provinces starts between Dominica and Martinique and extends southwest to between Curacao and Bonaire, demarking a triangle at the southeastern corner of the Caribbean Plate.

2. Within the northeastern Caribbean, material in the Guadeloupe-Dominica region split off from Puerto Rico. The fauna of this region has had a primarily vicariant origin (i.e., it consists of the descendents of populations that were on Proto-Guadeloupe at the time of its original fragmentation from Puerto Rico). The Guadeloupe region has itself served as the source for the biota on the younger islands north of Guadeloupe up to the Anegada Passage, and the propagules have come north from Guadeloupe by overwater dispersal, in accordance with the prevailing currents.

3. Within the southeastern triangle of the Caribbean, St. Lucia, La Blanquilla, and Bonaire are members of a former island arc that was upstream of the eastward-moving Caribbean Plate. Collision with this island arc resulted in the addition of St. Lucia between Martinique and St. Vincent, and pushed La Blanquilla and Bonaire down onto the Venezuelan shelf.

Concerning the larger picture of the entire Caribbean region, the biological data suggest that some geologic material now in the West Indies was in the Pacific during the Cretaceous and comprised a Pacific archipelago. The biogeographically distinct southeastern corner of the Caribbean may have been sutured to the Caribbean Plate at that time.

3.1 Systematics of Anolis

3.1.1 Early research

Using squamation characters (number, size, shape, texture, and arrangement of scales), Underwood [369] proposed that the anoles of the eastern Caribbean comprise three sets having the relationship diagramed in Figure 3.1.

As already discussed in Chapter 2, the northern Lesser Antilles, from the Anguilla Bank through Dominica, are populated by the bimaculatus group, which itself contains two series, the wattsi series and the bimaculatus series.

The wattsi series are small brown lizards that typically perch near the ground and on the leaf litter. Each population from this series co-occurs with a population

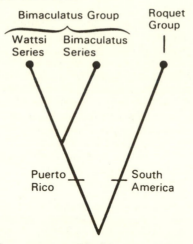

Fig. 3.1 Relationship among eastern Caribbean *Anolis* from early systematic research. The wattsi and bimaculatus series of the northern Lesser Antilles were presumed to represent separate waves of invasion from Puerto Rico, with the wattsi series being the more recent invasion. The roquet series of the southern Lesser Antilles was presumed to be derived from South America.

from the bimaculatus series; no population of the wattsi series occurs alone on an island.

The bimaculatus series consists of colorful large- and medium-sized lizards that are relatively arboreal. Some bimaculatus populations are alone on an island and others co-occur with a member of the wattsi series. Underwood [369] envisaged that each of these series represented a separate invasion from Puerto Rico and that the wattsi series was the more recent invasion.

The southern Lesser Antilles, from Martinique through Grenada, are populated by the roquet group: Today it is also known that the anoles on Bonaire in the Netherlands Antilles and La Blanquilla in the Venezuelan Antilles are also members of this group. The roquet group was presumed to have colonized from South America.

3.1.2 Contemporary studies

Systematic research since the species were originally described has clarified the phylogenetic relationships of the species and corroborates the early definitions of species groups based on squamation. The new data consist of karyotypes (descriptions of chromosomes), the immunological analysis of blood albumins, and the electrophoresis of proteins from over 20 loci [124, 123, 244, 395, 127, 126, 125, 331].

Figure 3.2 offers my synthesis of the phylogenetic relationships entailed by this recent work, based on the notes detailed in Tables 3.1A and 3.1B. The taxa being classified are the populations on each of the island banks in the Lesser Antilles, together with all the Puerto Rican anoles. Geographically differentiated varieties (subspecies) are diagrammed as vertical tick marks to indicate each named variety. Notice that the islands in the center of the arc—Guadeloupe, Dominica, and

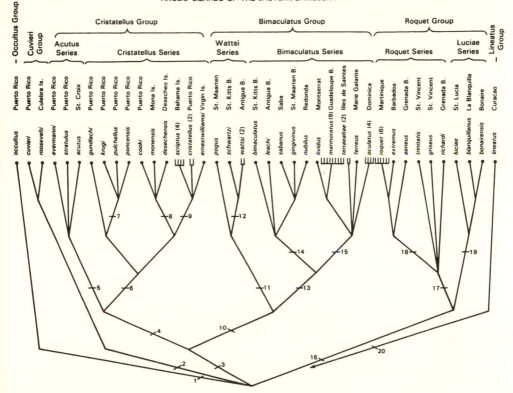

Fig. 3.2 Phylogenetic tree for *Anolis* in the eastern Caribbean. The tree is derived from data on squamation, karyotypes, the electrophoresis of proteins, and the immunogenetic assay of albumins. The taxa are populations on present-day island banks, irrespective of present nomenclatural status as species or subspecies. Taxa having geographical variation within an island bank that resulted in the naming of subspecies (geographical races) are indicated as horizontal lines with vertical tick marks to denote each subspecific name. The tree corroborates the early work in Figure 3.1. Also, the bimaculatus group is sister to the cristatellus group of Puerto Rico, and the roquet group is divided into the roquet series and the luciae series.

Table 3.1 Notes on phylogeny of eastern Caribbean *Anolis*.

0. Primitive anole karyotype = 12:24:0 (12 macrochromosomes, 24 microchromosomes, 0 sex chromosomes).

1. Twig anoles. Primitive karyotype, short prehensile tail, reduced lateral exposure of coronoid bone, AID(*evermanni → occultus*) = 49, AID(*extremus → occultus*) = 51.

2. Giant anoles. Primitive karyotype, AID(*evermanni → cuvieri*) = 45, AID(*extremus → cuvieri*) = 53, *A. roosevelti* is recently extinct on Culebra.

3. Central Caribbean complex. Derived karyotype: 14 or more macrochromosomes and sex chromosomes, AID(*evermanni → extremus*) = 59, AID(*cristate llus → extremus*) = 54, AID(*bimaculatus → extremus*) = 59, AID(*wattsi → extremus*) = 54.

4. Cristatellus group. AID(*evermanni → cristatellus*) = 16, AID(*evermanni → bimaculatus*) = 33, AID(*cristatellus → bimaculatus*) = 41.

5. Acutus series. Karyotype is 14 macrochromosomes, but other aspects of the karyotype differ among members of this series: *evermanni* = 14:10:2, *acutus* = 14:10:3, *stratulus* 14:12:3, AID's within the group to *evermanni* = 13-14. The Jamaican anoles also have 14 macrochromosomes; their AID's to *evermanni* = 27- 30, to *cristatellus* = 34-41, and to *bimaculatus* = 42-44. Thus the Jamaican anoles appear to be a clade of the acutus series. *A. acutus* is on St. Croix.

6. Cristatellus series. Each has 16 macrochromosomes, 8 to 10 microchromosomes, and 3 sex chromosomes. AID's within group to *cristatellus* = 11-13, Karyotypic, albumin, and electrophoretic data corroborate a division into a gundlachi line and a cristatellus line.

7. Grass anoles.

8. Trunk-ground anoles. Karyotype = 16:10:3, western Puerto Rico.

9. Trunk-ground anoles. Karyotype = 16:8:3, eastern Puerto Rico.

10. Bimaculatus group. Karyotype = 18:8:3, except *oculatus* of Dominica = 20:8:3. AID's within group to *bimaculatus* = 10-12.

11. Wattsi series. Small and brown with ventral scales strongly keeled. Division into wattsi series and bimaculatus series corroborated by electrophoretic data.

12. Genetic distance between *wattsi pogus* and other *wattsi* populations is greater than the distance between *w. schwartzi* and *w. wattsi*.

Table 3.1 Notes on phylogeny of eastern Caribbean *Anolis* (*continued*).

13. Bimaculatus series. Large or solitary-sized, often colorful, usually with green or grey-green tones. Division into northern and southern lines according to squamation and electrophoretic data.

14. Northern bimaculatus line. Ventral scales smooth. Two sublines: the large anoles – *A. bimaculatus bimaculatus* and *A. b. leachi*, frequently in upper canopy, and have large nuchal crests; medium-sized anoles *A. sabanus*, *A. gingivinus*, and *A. nubilus*, females with white lateral flank stripes; all three are electrophoretically closer to each other than are the two large anoles to each other.

15. Southern bimaculatus line. Ventral scales faintly keeled in at least some individuals. In spite of extensive differentiation in appearance as recognized by the subspecific nomenclature, electrophoretic distances among these taxa are slight. *A. lividus* from Montserrat and *A. marmoratus* of Guadeloupe are closest electrophoretically, *A. oculatus* of Dominica has karyotype = 20:8:3, and faintly keeled scales. *A. lividus* has the strongest keeling in the line, approaching, in some individuals, that of the wattsi series. The individuals from eastern half of Guadeloupe (Grande Terre) have smooth ventrals, and those from the western half (Basse Terre) are faintly keeled. *A. m. terraealtae* of the Ille des Saintes has smooth scales, and some females have white lateral flank stripes. *A. m. alliaceus* from forests on Basse Terre in Guadeloupe is similar in appearance to *A. bimaculatus* and was considered a subspecies of *A. bimaculatus* by Underwood [369] until more extensive collections showed it to be conspecific with all the other anoles on Guadeloupe.

16. Roquet group. Primitive karyotype, AID's to other lineages > 50, derived squamation on snout. Electrophoretic and albumin distances corroborate division into luciae and roquet series. AID's within roquet series to *extremus* = 0-14, and within luciae series to *extremus* = 26-35.

17. Roquet series. Four unresolved lines according to electrophoretic and albumin distance data. *A. griseus* and *A. richardi* are large, while *A. trinatatis* and the roquet line are medium-sized.

18. Roquet line. Derived karyotype = 12:22:0. Electrophoretic and albumin data corroborate that *extremus* is nearly identical to *roquet*. *A. aeneus* shows extensive clinal variation in appearance, but subspecies were not named.

19. Luciae series. Electrophoretic and albumin data corroborate the relative closeness of *blanquillanus* to *bonairensis*.

20. Lineatus series. *A. lineatus* has derived transverse processes on autotomic caudal vertebrae as do the anoles of Jamaica. AID(*lineatus* → *evermanni*) = 33, AID(*lineatus* → *cristatellus*) = 34, AID(*lineatus* → *valencienni* of Jamaica) = 22.

Martinique—have species with geographical variation (i.e., multiple subspecies); the species on the other islands show little internal differentiation. The major branches express a strict consensus among the morphological, karyotypic, and biochemical data.

The most important points are: (1) The distinction between the bimaculatus and roquet groups reflects a deeply different chromosomal makeup and great biochemical dissimilarity in blood albumins, (2) The bimaculatus group is sister to the cristatellus group of Puerto Rico, and both belong to a lineage termed the Central Caribbean Complex [331], (3) The genetic distances among the populations in the wattsi series are as great or greater than those in the bimaculatus series, (4) The roquet group subdivides into a roquet series and a luciae series, with the luciae series comprising the populations on St. Lucia, La Blanquilla, and Bonaire.

The distinction between the roquet and bimaculatus groups has proved so significant that the roquet group may deserve generic status, with the generic name *Dactyloa* [130].

Locating continental relatives of Caribbean anoles is difficult, even though over 150 species occur in Central and South America. The continental anoles are only distantly related to Caribbean anoles.

The closest known relative of an anole from the northeastern Caribbean, *A. evermanni* from Puerto Rico, is *A. gadovi*, from the Pacific [sic] versant of tropical Mexico. *A. gadovi* is itself a member of a cluster of rather closely related species in western Mexico, called the gadovi group. The albumin distance of the Puerto Rican anole to this Mexican anole is 29 units, a value far less than, for example, the typical albumin distance of about 55 to 60 units between a bimaculatus and roquet anole [128]. (See Tables 3.1A and 3.1B for other known albumin distances.) The karyotype of *A. gadovi* also includes sex chromosomes, which may be considered a derived characteristic shared with the northeastern Caribbean anoles, all of which also have sex chromosomes [187].

Perhaps even more surprising, the closest known non–West Indian relative of an anole from the southeastern Caribbean is *A. agassizi* on the isolated island of Malpelo in the Pacific south of Panama and west of Columbia [265]. The albumin distance between *A. extremus* of Barbados (itself nearly identical to *A. roquet* of Martinique) and *A. agassizi* of Malpelo is 40 units [331]. The distance of *A. extremus* to a typical anole of Venezuela, *A. chrysolepis*, is about 60, indicating a large distance. Thus, the preliminary indications are that the continental relatives of Caribbean anoles are on the Pacific coast of Central and northern South America.

The phylogenetic tree contributes to refuting the hypothesis that anoles colonize the Lesser Antilles by successive overwater dispersal from Puerto Rico in the north and from Venezuela in the south. By this hypothesis, the more recent invasions from a source region should still retain a close relationship to the taxa in the source region from which they were derived. In the north, the anoles are more closely related to the species on Guadeloupe at the center of the arc than to anything in Puerto Rico, and in the south, anoles are more closely related to the species on Martinique in the center of the arc than to anything in Venezuela. Thus, the sources for the Lesser Antillean anoles would appear to come from the center of the arc rather than from above or below, as traditionally thought.

Indeed, among the nine named varieties of *A. marmoratus* on Guadeloupe, one can find every distinguishing trait possessed by its relatives to the north, suggesting that some of the northern members of the bimaculatus series have been derived from Guadeloupe. Dispersal from the Guadeloupe region north is also consistent with the prevailing currents [272], whereas dispersal from Puerto Rico down into the Lesser Antilles is contrary to those currents.

Finally, the biochemical differentiation among the bimaculatus populations is appreciably less than that among the cristatellus populations or among the roquet populations, suggesting that the bimaculatus series is younger than the cristatellus or roquet series.

3.2 Biogeographic data

3.2.1 Nested-subset species/area relationship

The tiny cays on the Puerto Rico Bank (British Virgin Islands) demonstrate how a complex herpetofauna responds to habitat fragmentation. These cays reveal that the species lists for a sequence of islands of decreasing area comprise, approximately, a sequence of nested subsets. The cays are not random subsets of some larger fauna. Specifically, all the cays with only one species of amphibian or reptile have precisely *A. cristatellus*, those with two species have both *A. cristatellus* and a gecko *Sphaerodactylus macrolepis*, those with three or more species have these plus certain others. Each species seems to have a minimum island area at which it is first found and is present on all islands larger than its minimum area. Thus, the islands with only one species do not possess random species, nor do islands with only two species possess random pairs, drawn, as it were, from some urn of all species in the region. Figure 3.3 presents the data for the British Virgin Islands [181]. The lizards on cays of the Bahama Bank also form nested subsets [320].

During the Pleistocene, 15,000 years ago, the Puerto Rico Bank extending to Anegada was above water [138]. As the glaciers melted, slight hilltops became the tiny cays of the British Virgin Islands today. The lizards on these cays presumably did not have to disperse there; they were there to begin with. The fauna on the cays consists of special species that have the property of not becoming extinct when the size of their habitat shrinks. They have, so to speak, a good "bottle-necking" ability; colonizing ability is probably irrelevant to their presence on the cays of the Puerto Rico Bank.

A. cristatellus, the particular lizard that is retained on even the smallest of the Virgin Island cays, is a member of the sister taxon to the bimaculatus group of the northern Lesser Antilles. This fact supports an hypothesis that the fauna in the northern Lesser Antilles is a differentiated fragment extracted from the Puerto Rican fauna.

3.2.2 Amphibians and reptiles of Lesser Antilles

Table 3.2 (parts A-D) and Table 3.2 (parts a-d) present all the amphibians and reptiles of the Lesser Antilles today compiled from the checklist of Schwartz and Thomas [325] and the magnificent field guide of Schwartz and Henderson [324].

The faunal break at Dominica marking the end of the bimaculatus group of anoles in the Lesser Antilles is observed by *Ameiva* ground lizards, *Typhlops* blind snakes, and *Alsophis* snakes. The *Ameiva* in St. Vincent and Grenada in South American and the *Liophus* snakes of the southern islands are a South American genus. Also, Dominica is the southern end of the *Sphaerodactylus fantasticus* complex, and the *S. vincenti* complex begins in the adjacent Martinique where seven subspecies are found. It would be useful to know if the *S. vincenti* complex

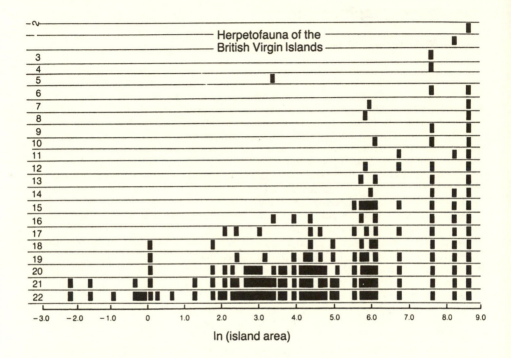

Fig. 3.3 Nested subset pattern for the relation between island area and species occurrence for amphibians and reptiles in the British Virgin Islands. Notice that *Anolis cristatellus* is the species that can survive on the smallest cay; it has the best "bottle-necking" ability, and presumably for this reason is the parental lineage to the bimaculatus group centered in the northern Lesser Antilles. The species corresponding to each of the numbers is: 1. *Eleutherodactylus cochranae*, 2. *Iguana pinguis*, 3. *Sphaerodactylus parthenopion*, 4. *Bufo lemur*, 5. *Thecadactylus rapicauda*, 6. *Eleutherodactylus antillensis*, 7. *Geochelone carbonaria*, 8. *Epicrates monensis*, 9. *Eleutherodactylus schwartzi*, 10. *Arrhyton exiguus*, 11. *Leptodactylus albilabris*, 12. *Amphisbaena fenestrata*, 13. *Iguana iguana*, 14. *Typhlops richardi*, 15. *Anolis pulchellus*, 16. *Alsophis portoricensis*, 17. *Mabuya sloanei*, 18. *Hemidactylus mabouia*, 19. *Anolis stratulus*, 20. *Ameiva exsul*, 21. *Sphaerodactylus macrolepis*, 22. *Anolis cristatellus*.

is very distantly related to the *S. fantasticus* complex, as are the anoles on these adjacent islands.

The Guadeloupe Archipelago is unique among the northern islands in having a great endemic differentiation of anoles, *Sphaerodactylus* geckos, *Alsophis* snakes, and *Eleutherodactylus* frogs. *A. marmoratus* is recognized as one species having 12 subspecies on Guadeloupe and the nearby islands, including the Ille de Saintes and Marie Galante. The species on Montserrat and Dominica are very close to *A. marmoratus*. Similarly, the *Sphaerodactylus* is recognized as one species having seven subspecies on Guadeloupe and nearby islands, together with subspecies on Montserrat and Dominica. The *Alsophis* is recognized as one species having two subspecies on Guadeloupe and nearby islands, together with subspecies on the Antigua bank, Montserrat, and Dominica. The *Eleutherodactylus* is recognized as four species, including two species that occur only in middle-elevation forest on Basse-Terre in Guadeloupe.

St. Lucia is as anomalous for other components of the herpetofauna as it is for *Anolis*. The closest relatives of *A. luciae* are the anoles of Bonaire and La Blanquilla near the coast of Venezuela, not the anoles from adjacent islands. St. Lucia is also anomalous in having the only hylid frog of the Lesser Antilles, *Hyla rubra* from South America; the only *Cnemidophorus* lizard of the Lesser Antilles, *C. vanzoi* similar to the widespread *C. lemniscatus* of South America; a South American gecko, *Hemidactylus palaichthus*; and a fere-de-lance, *Bothrops caribbaea*, nearly identical to the widespread South American *Bothrops atrox*.

Competition appears related to body size in some other lizards, as it is in *Anolis*. The two species of *Iguana* having the same body size show a perfect checkerboard distribution. Every island has an iguana, but the two species do not co-occur anywhere. Even on Guadeloupe *I. delicatissima* occurs on Grand Terre, while *I. iguana* is on Basse Terre. Similarly, no habitat has two species of *Ameiva*, all of which have about the same body size in the Lesser Antilles. *Ameiva pleei* is endemic to Sombrero and *A. corvina* occurs throughout the rest of the Anguilla bank. Yet on Hispaniola, where there are four species, the two extremes in body size coexist in the same habitat.

The Puerto Rico Bank has five species of *Amphisbaena* burrowing lizards. One species is found today in the British Virgin Islands on cays larger than 300 Ha. None, however, is found in the Lesser Antilles. The Puerto Rico Bank also has four species of *Typhlops* blind snakes, and one occurs in the British Virgin islands on cays larger than 350 ha. The Lesser Antilles do have a *Typhlops* on the St. Kitts and Antigua Banks through to Dominica.

3.2.3 Other taxa of West Indies

While it is beyond the scope of this chapter to review all known distributions of plants and animals in the Caribbean, the following paragraphs should help to place the data on amphibians and reptiles in perspective.

Perhaps the most intriguing biogeographic reference relative to today's tectonic setting is Alexander Agassiz's (1835–1910) "demonstration that the deep-water animals of the Caribbean Sea are more nearly related to those of the Pacific depths than they are to those of the Atlantic" (Singer [332, p. 255]). Agassiz concluded that the Caribbean was once a bay of the Pacific that since the Cretaceous has been cut off from the Pacific by the uplift of the Isthmus of Panama. Today, this observation would be taken as consistent with the hypothesis that much of the Caribbean sea floor was in the Pacific during the Cretaceous. The most recent

Table 3.2 **(Part A)** Herpetofauna of northern Lesser Antilles.

	Anguilla	Saba	St. Kitts
Bufo			*marinus*[A]
Hyla			
Leptodactylus			*fallax*[C]
Eleutherodactylus	*johnstoni*[E]	*johnstoni*	*johnstoni*
Mabuya	*mabouya*[K]		
Diploglossus			
Ameiva	*corvina*[M] *pleei*[N]		*erythrocephala*[O]
Cnemidophorus			
Bachia			
Kentropyx			
Gymnophthalmus			
Thecadactylus	*rapicauda*[1]	*rapicauda*	*rapicauda*
Hemidactylus	*mabouia*[2]	*mabouia*	*mabouia*
Phyllodactylus			
Sphaerodactylus	*sputator*[5] *macrolepis*[6]	*sabanus*[7]	*sputator* *sabanus*
Iguana	*delicatissima*[c]	*iguana*[d]	*delicatissima*
Anolis	*gingivinus*[e] *wattsi*[f]	*sabanus*[g]	*bimaculatus*[h] *wattsi*[i]
Typhlops			*monastus*[v]
Leptotyphlops			
Bothrops			
Boa			
Corallus			
Alsophis	*rijersmai*[gg]	*rufiventris*[hh]	*rufiventris*
Liophus			
Chironius			
Clelia			
Mastigodryas			
Pseudoboa			

Table 3.2 (Part A) Herpetofauna of northern Lesser Antilles (*continued*).

Antigua	Montserrat	Guadeloupe	Dominica
marinus	*marinus*	*marinus*	
fallax	*fallax*	*fallax*	
johnstoni	*johnstoni*	*johnstoni*	
martinicensis[F]		*martinicensis*	*martinicensis*
	barlagnei[G]		
		pinchoni[H]	
griswoldi[P]	*mabouya*	*mabouya*	*mabouya*
	montisserrati[L]		
	pluvianotata[Q]	*cineracea*[R]	*fuscata*[S]
rapicauda		*underwoodi*[X]	
mabouia	*rapicauda*	*rapicauda*	*rapicauda*
	mabouia	*mabouia*	*mabouia*
elegantulus[8]			
	fantasticus	*fantasticus* spp.[9]	*fantasticus*
delicatissima			
		delicatissima	*delicatissima*
bimaculatus[j]	*iguana*	*iguana*	
wattsi spp.[k]	*lividus*[l]	*marmoratus* spp.[m]	*oculatus* spp.[n]
monastus	*monastus*[w]	*dominicana*[x]	*dominicana*[y]
			constrictor[dd]
antillensis	*antillensis*	*antillensis* spp.[ii]	*antillensis*
		juliae spp.[ij]	*juliae*
		clelia[pp]	

Table 3.2 (Part B) Herpetofauna of southern Lesser Antilles.

	Martinique	St. Lucia
Bufo	*marinus*	*marinus*
Hyla		*rubra*[B]
Leptodactylus		*fallax*
Eleutherodactylus	*johnstoni*	*johnstoni*
	martinicensis	
		mabouya
Mabuya		
Diploglossus		*vanzoi*[U]
Ameiva		
Cnemidophorus		
Bachia		*pleei* spp.[Z]
Kentropyx		*rapicauda*
Gymnophthalmus	*pleei*[Y]	*mabouia*
Thecadactylus	*rapicauda*	*palaichthus*[3]
Hemidactylus	*mabouia*	
Phyllodactylus		*microlepis* spp.[b]
Sphaerodactylus		*vincenti*
	vincenti spp.[a]	
Iguana	*delicatissima*	*iguana*
		luciae[p]
Anolis	*roquet* spp.[o]	
Typhlops		
Leptotyphlops	*bilineata*[aa]	*bilineata*
Bothrops	*lanceolata*[bb]	*caribbaea*[cc]
Boa		*constrictor*[ee]
Corallus		
Alsophis		
Liophus	*cursor*[kk]	*ornatus*[ll]
Chironius		
Clelia		*clelia*
Mastigodryas		
Pseudoboa		

Table 3.2 (Part B) Herpetofauna of southern Lesser Antilles (*continued*).

St. Vincent	Grenada	Barbados
marinus	*marinus*	*marinus*
wagneri[D]	*wagneri*	
johnstoni	*johnstoni*	*johnstoni*
urichi[l]	*urichi*[J]	
mabouya	*mabouya*	*mabouya*
ameiva[T]	*ameiva*	
	heteropus[V]	
		copei[W]
underwoodi		*underwoodi*
rapicauda	*rapicauda*	
mabouia	*mabouia*	
		pulcher[A]
vincenti		
iguana	*iguana*	
griseus[q]	*richardi*[s]	*extremus*[u]
trinatatis[r] *aeneus*[t]		
	tasymicris[Z]	
		bilineata
enydris[ff]	*enydris*	
	melanotus[mm]	*perfuscus*[nn]
vincenti[oo]		
	clelia	
bruesi[qq]	*bruesi*	
	neuwiedi[rr]	

Table 3.2 Notes on herpetofauna of Lesser Antilles.

A. *Bufo marinus*. Large toad tolerant of brackish conditions; also in Greater Antilles, except Cuba (where two large species of *Bufo* are endemic) and Central and South America.

B. *Hyla rubra*. Also Central and South America.

C. *Leptodactylus fallax*. Endemic in Lesser Antilles, recently extinct on St. Kitts, Guadeloupe, and St. Lucia.

D. *Leptodactylus wagneri*. Also South America.

E. *Eleutherodactylus johnstoni*. Endemic in Lesser Antilles.

F. *Eleutherodactylus martinicensis*. Endemic in Lesser Antilles, also on La Desirade, and Ille des Saintes.

G. *Eleutherodactylus barlagnei*. Endemic to Guadeloupe, 600-2100 ft. elevation on Basse Terre.

H. *Eleutherodactylus pinchoni*. Endemic to Guadeloupe, 600-2200 ft. elevation on Basse-Terre.

I. *Eleutherodactylus urichi schrevei*. Subspecies endemic to St. Vincent, *E. u. urichi* in Trinidad and Venezuela.

J. *Eleutherodactylus urichi euphronides*. Subspecies endemic to Grenada.

K. *Mabuya mabouya mabouya*. Also in Trinidad, Tobago, and South America.

L. *Diploglossus montisserrati*. Endemic to Montserrat. Genus occurs on Greater Antilles, except Jamaica.

M. *Ameiva corvina*. Endemic to Sombrero.

N. *Ameiva pleei*. Endemic to Anguilla Bank.

O. *Ameiva erythrocephala*. Endemic to St. Kitts Bank.

P. *Ameiva griswoldi*. Endemic to Antigua Bank.

Q. *Ameiva pluvianotata*. Endemic to Lesser Antilles, with distinct subspecies on Montserrat and Redonda.

R. *Ameiva cineracea*. Endemic to Guadeloupe, known only from Grand Ilet, off east coast of Basse-Terre, now extinct.

S. *Ameiva fuscata*. Endemic to Dominica. A specimen named "*A. major*" collected in 1839 may represent an extinct population from Martinique.

T. *Ameiva ameiva tobagana*. Subspecies endemic to St. Vincent and Grenada Bank, nominate subspecies occurs in Surinam.

Table 3.2 Notes on herpetofauna of Lesser Antilles (*continued*).

U. *Cnemidophorus vanzoi.* Endemic to Maria Islands off St. Lucia.

V. *Bachia heteropus alleni.* Endemic subspecies to Grenada Bank, other subspecies on Trinidad and Venezuela.

W. *Kentropyx copei.* Endemic to Barbados, only Antillean representative of a genus widespread in South America.

X. *Gymnophthalmus underwoodi.* Also found in Surinam.

Y. *Gymnophthalmus pleei pleei.* Endemic subspecies to Martinique.

Z. *Gymnophthalmus pleei.* Distinct subspecies endemic to St. Lucia and to the Maria Islands off St. Lucia, species endemic to Martinique and St. Lucia.

1. *Thecadactylus rapicauda.* Also in Trinidad, Tobago, South and Central America, not in Greater Antilles, monotypic genus.

2. *Hemidactylus mabouia.* Also in Africa, Madagascar, eastern South America, Trinidad and Tobago, Greater Antilles and Virgin Islands, except Jamaica.

3. *Hemidactylus palaichthus.* On Maria Islands off St. Lucia, also in South and Central America, and Trinidad and Tobago.

4. *Phyllodactylus pulcher.* Endemic to Barbados.

5. *Sphaerodactylus sputator.* Endemic to Anguilla Bank and St. Kitts Bank.

6. *Sphaerodactylus macrolepis parvus.* Subspecies endemic to Anguilla Bank, other subspecies on Puerto Rico Bank.

7. *Sphaerodactylus sabanus.* Endemic to Saba and St. Kitts Bank.

8. *Sphaerodactylus elegantulus.* Endemic to Antigua Bank.

9. *Sphaerodactylus fantasticus.* Nine subspecies in total, two on Basse-Terre and two on Grand Terre in Guadeloupe, and one apiece on Ille des Saintes, Marie Galante, La Desirade, Montserrat, and Dominica.

a. *Sphaerodactylus vincenti.* Nine subspecies in total, seven on Martinique, and one apiece on St. Vincent and St. Lucia.

b. *Sphaerodactylus microlepis.* Endemic to St. Lucia, with distinct subspecies on St. Lucia and the Maria Islands off St. Lucia.

c. *Iguana delicatissima.* Endemic to Lesser Antilles, including Grande Terre in Guadeloupe, La Desirade, and Terre-de-Bas and Terre-de-Haute in the Illes des Saintes.

d. *Iguana iguana.* Also South America, also Lesser Antilles and Virgin Islands including Basse-Terre in Guadeloupe; La Coche, Grand Ilet, Terre-de-Haute, and Ilet a Cabrit in the Illes des Saintes; and the Maria Islands off St. Lucia.

Table 3.2 Notes on herpetofauna of Lesser Antilles (*continued*).

e. *Anolis gingivinus*. Endemic to Anguilla Bank.

f. *Anolis wattsi pogus*. Subspecies endemic to Anguilla Bank.

g. *Anolis sabanus*. Endemic to Saba.

h. *Anolis bimaculatus bimaculatus*. Subspecies endemic to St. Kitts Bank, species endemic to Lesser Antilles.

i. *Anolis wattsi schwartzi*. Subspecies endemic to St. Kitts Bank.

j. *Anolis bimaculatus leachi*. Subspecies endemic to Antigua Bank.

k. *Anolis wattsi*. Species endemic to Lesser Antilles, distinct subspecies on Barbuda and Antigua.

l. *Anolis lividus*. Endemic to Montserrat.

m. *Anolis marmoratus*. Twelve subspecies total, two on Grand Terre and four on Basse-Terre in Guadeloupe, two in the Illes des Saintes, and one apiece on Ilet-a-Kohouanne, La Desirade, Les Iles de la Petite Terre, and Marie Galante.

n. *Anolis oculatus*. Endemic to Dominica with four subspecies.

o. *Anolis roquet*. Endemic to Martinique with six subspecies.

p. *Anolis luciae*. Endemic to St. Lucia.

q. *Anolis griseus*. Endemic to St. Vincent.

r. *Anolis trinatatis*. Endemic to St. Vincent.

s. *Anolis richardi*. Endemic to Grenada Bank.

t. *Anolis aeneus*. Endemic to Grenada Bank.

u. *Anolis extremus*. Endemic to Barbados.

v. *Typhlops monastus geotomus*. Subspecies endemic to Antigua and St. Kitts Banks.

w. *Typhlops monastus monastus*. Species endemic to Lesser Antilles, subspecies endemic to Montserrat.

x. *Typhlops dominicana guadeloupensis*. Subspecies endemic to Guadeloupe, known only from Grande-Terre.

y. *Typhlops dominicana dominicana*. Species endemic to Lesser Antilles, subspecies endemic to Dominica.

z. *Typhlops tasymicris*. Endemic to Grenada Bank, presumed close to South American forms.

Table 3.2 Notes on herpetofauna of Lesser Antilles (*continued*).

---

aa. *Leptotyphlops bilineata.* Endemic to Lesser Antilles.

bb. *Bothrops lanceolata.* Endemic to Martinique, genus widespread in South America.

cc. *Bothrops caribbaea.* Endemic to St. Lucia, presumed close to *B. atrox* of South America.

dd. *Boa constrictor nebulosa.* Subspecies endemic to Dominica.

ee. *Boa constrictor crophias.* Subspecies endemic to St. Lucia.

ff. *Corallus enydris cooki.* Also South America, nominate subspecies in Amazonian region.

gg. *Alsophis rijersmai.* Endemic to Anguilla Bank.

hh. *Alsophis rufiventris.* Endemic to Saba and St. Kitts Bank.

ii. *Alsophis antillensis.* Five subspecies total, one apiece on the Antigua Bank, Montserrat, Guadeloupe, Marie Galante, Ille des Saintes, and Dominica.

jj. *Liophus juliae.* Three subspecies total, one apiece on Guadeloupe, Marie Galante, and Dominica.

kk. *Liophus cursor.* Endemic to Martinique.

ll. *Liophus ornatus.* Endemic to St. Lucia, collected on Maria Islands off St. Lucia.

mm. *Liophus melanotus.* Also Trinidad, Tobago, and South America.

nn. *Liophus perfuscus.* Endemic to Barbados.

oo. *Chironius vincenti.* Endemic to St. Vincent; a similar form, *C. carinatus*, occurs in South and Central America and Trinidad.

pp. *Clelia clelia.* Also South and Central America.

qq. *Mastigodryas bruesi.* Endemic to St. Vincent and Grenada Bank, genus in South America.

rr. *Pseudoboa neuwiedi.* Also in South and Central America.

---

analysis of tectonic implications for marine biogeography in the Caribbean is offered in [101].

Ten genera of plants from the Greater Antilles that reach the northern Lesser Antilles [149, Table VI, p. 23]. Of these, five terminate at Guadeloupe and the rest north of Guadeloupe; that is, Guadeloupe is as far south as these genera reach.

Schwartz [323] reviewed the distribution of amphibians and reptiles throughout the Greater and Lesser Antilles. Table 3.2A for the Lesser Antilles is drawn from this and his later work. For the Greater Antilles, highlights include the existence of endemic satellite genera to the anoles: *Chamaeleolis* with three species on Cuba, and *Chamaelinorops* from the south island of Hispaniola. The prehensile tail and general aspect of *Chamaeleolis* suggests an African chamaeleon. The Chamaeleonidae and Agamidae together form the sister lineage to the Iguanidae, with a split estimated at mid-Jurassic [106]. Cuba also has an endemic genus of burrowing lizard, *Cadea*, with two species. And of special interest, Cuba has an endemic genus, *Cricosaura*, of the family Xantusiidae, the night lizards (cf. [135]). The other three genera of this family occur in the southwestern United States, on the Channel Islands off southern California, and in central Mexico and Central America.

West Indian mammal and bird fossils come primarily from the Pleistocene and Holocene [234]. Extinct faunas in the Greater Antilles are generally rodents, insectivores, and edentates, together with raptorial birds, often of great size that presumably evolved in response to the lack of mammalian predators. Moreover, a dramatic bear-sized rodent, *Amblyrhiza inundata*, in an endemic Antillean family, Heptaxodontidae, was excavated from mines on Anguilla [76, 216]. Olson concludes that this fauna is the result of multiple overwater colonizations in the Oligocene and later, with the mammals having stronger affinities with South America and the birds with North America. These data indicate the limited value of mammals for tectonic reconstruction of the Caribbean. Mammals have been available as colonists only since the mid-Tertiary after much of the tectonic events that created the Caribbean had already occurred.

For bats, as with reptiles and amphibians, Guadeloupe is a center of endemism; 6 of 10 bat species are endemic to the Antilles [18]. A significance to overwater dispersal seems obvious. In contrast with anoles, 11 species of bats in the southern Lesser Antilles show a classic distance effect with respect to South America, as illustrated in Reference [18, Figure 2, p. 77].

The freshwater fish of the Greater Antilles are secondary, meaning that they are derived from salt-water ancestors [41]. Also, the diversity of endemic genera and species of freshwater fishes is highest in Cuba and declines progressively eastward. No endemic freshwater fish occur in the Lesser Antilles and therefore can contribute only limited information about the origin of the eastern Caribbean.

A suborder (Auchenorrhyncha) of the Homoptera for the Greater Antilles is derived primarily from Central America and Mexico, with little or no affinity to North or South America [262]. Ramos [262] emphasized agreement with Metcalf [220], who proposed defining a zoogeographic region, the Caribbean, to include the West Indies, Mexico, and Central America, and distinguished from the neotropical region.

The Scaritinae, an old and cosmopolitan group of Carabid burrowing ground beetles that consists of over 1500 species throughout the world, has single-island endemics in the Greater Antilles, Guadeloupe, and Martinique [231]. An endemic genus, *Antilliscaris*, found on Puerto Rico and Hispaniola, is closely aligned with genera from the Afrotropical region on the basis of larval characteristics. The presence of three species in this genus on Puerto Rico, together with a 38% endemism in the fauna on Puerto Rico, led Nichols to suggest that Puerto Rico is the oldest emergent landmass in the West Indies.

Another large group of Carabid beetles, the genus *Platynus*, has over 300 species in Mexico and Central America and 1000 or more worldwide. Most are found at elevations over 300 m. Many endemic species are found in the Greater Antilles, although, curiously, none occurs in Puerto Rico [188]. Five endemic species occur on Guadeloupe, four on Dominica, and one each on Martinique, St. Lucia, and St. Vincent.

Fifty-eight species of drosophilid flies are endemic to the Antilles [129]. Also, an endemic genus of drosophilid in the Caribbean is *Mayagueza*; it has only one species and is endemic to Puerto Rico. Moreover, the *Drosophila repleta* group that feeds primarily on decaying cactus has *D. peninsularis* endemic in the Greater Antilles; it is related to Central American stocks.

The classic power law of island biogeography relating number of species to island area is valid in the West Indies for ant species [390]. Curiously, no endemics are reported for the northern Lesser Antilles, including Guadeloupe and Martinique, but endemic species are reported from St. Lucia, St. Vincent, Grenada, and Barbados. This suggests the need for further collection and analysis in the Lesser Antilles. On the Greater Antilles a wealth of fossil ants are preserved in Tertiary amber from the Dominican Republic. The late Tertiary fauna of the Greater Antilles is more characteristic of a continental fauna. Several genera were present on Hispaniola, including army ants, that are now absent from the West Indies but that remain abundant on the continent.

Other insect groups recently reviewed include the Lygaeidae where overwater dispersal is of primary importance [333]. There is limited endemism in the Lesser Antilles (only 1 in about 40 species on each of St. Kitts, Dominica, St. Vincent, and Grenada) and a distance effect with respect to South America. Although the Lygaeids are a Lower Cretaceous group, their biogeography in the Lesser Antilles seems rather recent. In the Greater Antilles, however, 119 species are endemic to the islands, including 4 endemic genera, 2 of which are in Cuba, 1 in Hispaniola, and the other more widely distributed in the West Indies. Similarly, the sweat bees (Halictidae) have also been interpreted primarily in terms of overwater dispersal in the Lesser Antilles [103]. Both these groups seem to accord with the butterflys, which have long been interpreted in terms of overwater dispersal [43].

As a generalization, then, Cuba and, perhaps to a lesser extent, Puerto Rico are old biologically, as evidenced by the endemic genera on these islands. The central islands of the Lesser Antilles are also places of endemism at the specific and subspecific level in many groups.

The biogeography within the Bahamas, a large carbonate platform with a continental basement, primarily reflects sea-level changes [118, 136].

Finally, the Antilles have long been conjectured as possibly providing land bridges between North and South America while they were west of their present positions closer to the continents. This issue is explored in detail in a marvelous volume edited by Stehli and Webb [358]. Recently, Perfit and Williams [245] have concluded, based on mammalian data, that an isthmus of continuous dry land did not exist between North and South America during the latest Cretaceous and earliest Tertiary (Paleocene). Yet, Stehli and Webb [359] have documented substantial evidence that mammals, reptiles, nonmarine mollusks, and angiosperm plants *did* filter in both directions between North and South America at this time. They envision island-arc stepping stones as the route of dispersal, not a continuous dry-land connection.

3.3 Geologic data

3.3.1 Faults between north and south

The phylogenetic tree of *Anolis* shows a major dichotomy between the cristatellus-bimaculatus lineage of Puerto Rico and the northern Lesser Antilles versus the roquet group of the southeastern triangle of the Caribbean region.

The deepest seismicity in the Lesser Antilles is found at the top of this triangle, between Dominica and Martinique [367, 370, 93]. The seismicity occurs there with an East-West trend that shows strike slip faulting at depth [360]. This fault is labeled as the Dominica Fault in Figure 3.4.

A line extending southwest from the fault between Dominica and Martinique to some point between Bonaire and Curacao appears to coincide with a major difference in the crustal structure of the ocean floor in the Venezuelan Basin. Three parallel faults trending east northeast termed the "central Venezuelan Basin fault zone" [27, 89] are illustrated in Figure 3.4. West of the central Venezuelan Basin fault the B" reflecting layer is smooth, as characteristic of most of the Caribbean, whereas it is described as rough east of this fault in the southeastern corner of the Caribbean. Moreover, this southeastern corner is a magnetically quiet zone, in contrast to the rest of the Venezuelan Basin where northeast-southwest trending magnetic anomalies parallel the central Venezuelan Basin Fault [119].

Bouysse [32] has suggested that the northern Lesser Antilles has developed separately from the southern Lesser Antilles, and Speed [338] has alluded to a similar possibility. The data on *Anolis* strongly support this kind of hypothesis. Presumably, this southeastern triangle, which harbors the roquet group of *Anolis*, is a small separate plate or a piece of the South American Plate that has become sutured to the present-day eastward-moving Caribbean plate.

3.3.2 Guadeloupe derived from Puerto Rico

Now focus on the northeastern part of Lesser Antilles. La Desirade, on the Guadeloupe Archipelago, has Jurassic rock [113]. After repeated study, no doubt remains as to its magmatic origin, similar in appearance, chemical composition, and mineralogy to the basal units of the Greater Antilles, and especially Puerto Rico [35]. Although a tiny sliver of land on an aerial map (ca 30 km^2), La Desirade is the tip of an "iceberg," marking the edge of a cliff that drops 5000 m to the ocean floor in less than 10 km horizontal distance, and is the site of a distinct local positive gravity anomaly (Bourguer anomaly). Tomblin [367, p. 471] wrote ... in view of the long time interval between these and the next oldest rocks in the Lesser Antilles and the large horizontal movements which undoubtedly took place in the region during this interval ... it is possible that the crust now exposed in Desirade represents a small block which was much closer to the Virgin Islands in Late Jurassic time and which subsequently separated from this area and moved relatively eastward. Alternatively, basement as old as that exposed on La Desirade may underly all the northern Lesser Antilles but simply not be exposed between the Guadeloupe region and Puerto Rico. Many authors [31, 32, 338, 92] have referred to the origin of the northern Lesser Antilles out to Guadeloupe as occurring together, in some way, with Puerto Rico.

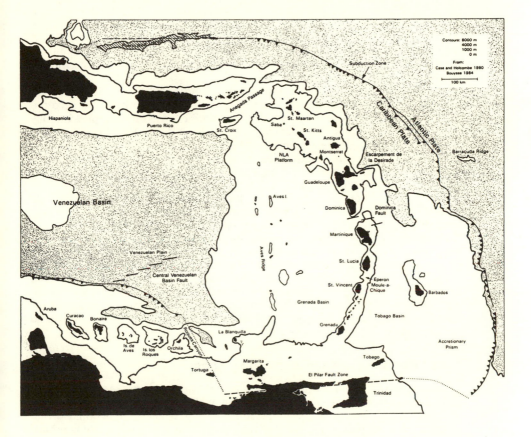

Fig. 3.4 Tectonic map of Lesser Antilles. Notice the 1-km bathymetric countour. The bimaculatus group occupies the northern Lesser Antillean Platform and is sister to the cristatellus group of Puerto Rico. The roquet group occupies the Lesser Antilles from Martinique south and extending to Bonaire.

The phylogenetic tree of *Anolis* supports an hypothesis that the Guadeloupe region has moved apart from the Puerto Rico Bank because the bimaculatus group of the northern Lesser Antilles is sister to the cristatellus group of Puerto Rico. The bifurcation in the phylogenetic tree where these lineages split would seem to coincide with the movement of a small block representing proto-Guadeloupe away from the Puerto Rican area. This block would carry on it the ancestor of *A. cristatellus*, just as the small cays of the Puerto Rico bank today retain *A. cristatellus* itself. This ancestor of Puerto Rico's *A. cristatellus* developed in isolation on proto-Guadeloupe to become the present-day *A. marmoratus*. And this *A. marmoratus*, in turn, spawned propagules that dispersed north, with the prevailing currents, to islands younger than Guadeloupe as they appeared. In fact, the entire range of the bimaculatus group coincides exactly with the geologic entity bounded by the 1000-m-depth contour encompassing the area from the Anegada Passage to the passage between Dominica and Martinique.

The Anegada Passage between the Virgin Islands and the northern Lesser Antilles is an extensional basin today, and perhaps this type of expansion has occurred before. Alternatively, the Aves Ridge, when more exposed than at present, may have provided a stepping-stone path for dispersal from Puerto Rico to Guadeloupe, assuming that Guadeloupe was emergent before the islands north of it. This alternative would not require an expansion of the arc between Guadeloupe and Puerto Rico. Still, migration down an exposed Aves Ridge to Guadeloupe would run counter to the ocean currents, and such migration would have had to involve many species to account for the large fauna on Guadeloupe today. Thus, I feel the data favor an expansion of the arc between Guadeloupe and Puerto Rico.

I term the structure that contains the present-day islands of the Anguilla Bank down through Dominica as the "NLA platform." Faunal buildup involving an invasion, coevolution, and possible extinction of *Anolis* species seems to be occurring on the Anguilla, Antigua, and St. Kitts Banks toward the northern part of this platform as a result of dispersal from the Guadeloupe Archipelago, as discussed in Chapter 2. The presence of wattsi-series anoles as the smaller species in these banks invites the conjecture that they were united at one time and colonized in common from Guadeloupe. After this megabank broke into the Anguilla, Antigua, and St. Kitts Banks, they were colonized a second time.

3.3.3   Bonaire, La Blanquilla, and St. Lucia

Turning now to the southeastern triangle of the Caribbean one sees that there is further heterogeneity. The coast of Venezuela, including the Netherlands-Venezuelan Antilles, appears to represent the remnants of an island arc that collided with the coast of South America in the late Cretaceous [208]. Curacao and Bonaire have different basements, formed at a mid-ocean spreading ridge and at a subduction zone, respectively [21, 22], and the anoles on these adjacent islands are not at all closely related. And in the Lesser Antilles, the chemical composition of the rocks on St. Lucia differ markedly from those of its neighbors [367, Figure 2], just as St. Lucia's anole is not closely related to those of its neighbors.

Bucher [47] suggested that the motion of the Caribbean Plate between the North and South American Plates is analogous to the motion of a glacier in a valley. The analogy explains the opposite symmetry of pull-apart basins with the strike-slip motion at faults in the north and south boundary zones of the Caribbean Plate [205]. Perhaps the analogy can be extended by viewing the southern Lesser Antilles as the "terminal moraine" at the southern half of the leading edge of the Caribbean

Plate. As this part of the Caribbean Plate has moved eastward, it accumulated some islands in its path. Specifically, St. Lucia (perhaps near Eperon Moule-a-Chique) may have a different geologic origin from its present-day neighbors of Martinique and St. Vincent and, instead, may have originated with Bonaire and La Blanquilla, both of which are now lodged on the Venezuelan coast.

3.3.4 Schematic of Lesser Antilles

Figure 3.5 presents a reconstruction of the geologic origin of the eastern Caribbean that takes into account the points raised above. The two halves of the Lesser Antilles are hypothesized to develop separately. In the north, Puerto Rico and Desirade split. As they move apart, the NLA platform, containing the present-day Anguilla, Antigua, and St. Kitts banks, grows in area. Thereafter the *A. wattsi* lineage and then the *A. bimaculatus* lineage colonize the exposed areas on the platform by overwater dispersal. In the south, matters are more complex. An island arc consisting of basal elements of Bonaire, La Blanquilla, and St. Lucia is hypothesized to lie in the path of the material that contributes to Martinique and St. Vincent–Grenada. As the Caribbean plates moves east, it pushes the islands between Bonaire and La Blanquilla onto the Venezuelan shelf and picks up St. Lucia.

3.4 Discussion

3.4.1 An ancient and heterogeneous Antilles

Tectonic hypotheses for the origin of the Caribbean Plate—why it exists at all and where it came from—are becoming increasingly refined, and the Caribbean region as a whole has recently received a magnificent review in Volume H of the *Decade of North American Geology* series (DNAG) published by the Geological Society of America [85]. Briefly, relevant to later discussion, here are some of the key points about how the Caribbean Plate formed.

North and South America separated during the late Jurassic, and by the early Cretaceous the sea between the Americas, called the "proto-Caribbean seaway," contained a spreading ridge. Also during the early Cretaceous, the Greater Antilles started to form in the eastern Pacific near the junction of the American and Farallon Plates. Next, during the middle Cretaceous, basalts were extruded onto pre-existing oceanic crust west of the proto–Greater Antilles. This resulted in a relatively thick buoyant lithosphere with a thickness of 15 to 25 km that can resist subduction [53, 250]. The initiation of the Galapagos hot spot apparently caused this basaltic extrusion, and crust now in the Caribbean was above the Galapagos hot spot at that time [97, 336, 98]. Thus, the situation is one of a proto–Greater Antilles lying in the Pacific, with an unusually light and thick lithosphere behind it, and this lithosphere together with the proto–Greater Antilles thereafter comprised the Caribbean Plate. Because it was not subductable, the Caribbean Plate passed between the westward moving American Plates and came to rest with the Greater Antilles located near the Bahama platform. While the Caribbean Plate moved between the Americas, it overrode the spreading ridge in the proto-Caribbean seaway, thereby erasing the features that would have revealed more exactly how the Americas originally split from each other. Finally, in the Eocene, the Caribbean Plate motion changed its direction from northeast to east, resulting in the Lesser Antilles becoming the

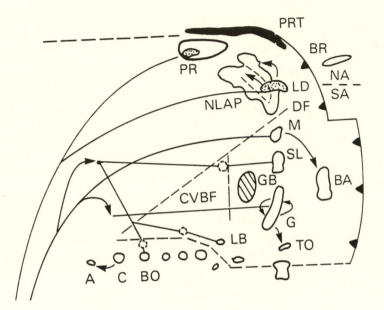

Fig. 3.5 Schematic for proposed geologic origin of the eastern Caribbean, together with historical biogeography of *Anolis*. Abbreviations are: PR, Puerto Rico; PRT, Puerto Rico Trench; BR, Barracuda Ridge; LD, La Desirade; NLAP, Northern Lesser Antilles Platform; DF, Dominica Fault; M, Martinique; SL, St. Lucia; BA, Barbados; GB, Grenada Basin; G, Grenada-St. Vincent; TO, Tobago; CVBF, Central Venezuelan Basin Fault; LB, La Blanquilla; BO, Bonaire; C, Curacao; A, Aruba. Lines ending in an arrow indicate overwater dispersal by anoles. Lines without arrows indicate the path traced by *Anolis* populations as they occupied geologic structures.

leading edge of the plate, while the Greater Antilles became the northern margin of the plate.

Concerning the Greater Antilles [184], western and central Cuba are mainly underlain by continental crust, with lower Jurassic formations extant in western Cuba and Precambrian material in central Cuba [271]. Following its Eocene collision with the Bahama platform, Cuba developed into a series of uplifted blocks separated by marine basins and presumably comprised an archipelago at this time. Throughout the Miocene, Cuba gradually emerged to assume its present above-water area by the Pliocene. In Hispaniola the oldest formations are early Cretaceous. Its present physiography results from accelerated uplift during the Piocene-Pleistocene, presumably as a result of motion along a restraining bend along the northern margin of the plate [206]. Puerto Rico has an Upper Jurassic and Lower Cretaceous material in the Bermeja complex toward the southeastern corner of the island. In Jamaica the oldest deposits are Cretaceous, but the island was completely immersed in the Eocene, only to emerge gradually in the early Miocene. Jamaica was an archipelago during the middle Miocene and fully emerged by the Pliocene. The ages of the oldest material known from the Greater Antilles is consistent with biogeographic data discussed above showing the greatest generic endemism in Cuba and Puerto Rico.

Concerning the Lesser Antilles, panoramic reconstructions of the Caribbean's origin generally view the Lesser Antillean island arc as (1) a homogeneous grouping of islands and (2) as forming during the Tertiary, e.g., [249, 250]. Yet new discoveries point to an older age for the basement of the Lesser Antilles than previously believed. In 1975, the oldest material known in the Lesser Antilles, apart from the Aves Ridge [251, 33] and then-controversial La Desirade find [113], was Eocene (ca 50 Ma), with large deposits on the NLA platform and on the Grenada Bank. These facts contributed to the acceptance of a Tertiary origin for the Lesser Antilles. Now, Upper Cretaceous (ca. 70 Ma) basement material has been obtained from both slopes of the Anegada Passage [339, 34, 32] and on Union Island in the Grenadines [376, 377]. These data show that the Lesser Antillean island arc is older than previously thought.

Furthermore, the geologic data are also not consistent with viewing the Lesser Antillean arc as a homogeneous group of islands. The isolated occurrence of Jurassic material in La Desirade is one example of heterogeneity. The difference in earthquake frequencies between the northern and southern halves of the arc, the occurrence of distinctive ocean floor in the southeast corner of the Caribbean, the difference between the basements of Curaçao and Bonaire, together with suggestions of a different lithic chemistry on St. Lucia, offer other examples. Biologically, the contrast between the bimaculatus anoles of the northern Lesser Antilles and the roquet anoles of the southern Lesser Antilles is stark, and the difference between the lineages of the anole on Curaçao and that on the adjacent island of Bonaire is almost as dramatic. Thus, many data point to a heterogeneous history for the structures that are now aligned at the leading edge of the Caribbean Plate.

3.4.2 A Pacific archipelago

Vertebrate paleontology and biogeography in Central America have focussed on the study of mammals as indicators of when the isthmus of Panama closed. This application of biogeography relies on the fact that mammal faunas of North and South America are distinctive, while Central American mammals are essentially a North American fauna. The closure of the isthmus is then indicated by dispersal of South American forms northward into Central and then North America, and North American forms into South America. This entire investigation has, however, diverted attention away a fundamental clue offered by the biogeography of other vertebrate groups. For amphibians, reptiles, and freshwater fish, Central America is a faunal province of its own, with many endemic genera, that have southern, not northern, affinity [54, 95, 96, 301]. To quote from Savage [303]: The distributions of recent and fossil mammals for the region have been extensively reviewed by several workers, most recently by Savage (1974) [302], Ferrusquia-Villafranca (1978) [112], and Marshall et al. (1979) [209]. These studies all confirm that the South American mammal fauna was isolated from that of Central America until Pliocene; that no distinctive Middle American mammal fauna can be recognized; that no cluster of taxa of southern relationships, equivalent to the Middle American unit seen in the freshwater fishes and herpetofauna, can be distinguished; and that the region was dominated by groups of northern affinity until the interchange with South America. It should also be noted that the Antillean herpeto-fauna is generally allied with Central America, and to some extent South America; for example, it does not contain the North American forms readily observed in the southwestern U.S.

These considerations led Savage [303] to propose that: There was a continuous land connection or series of proximate islands extending from northern South America to the area of Nicaragua in late Mesozoic and/or early Tertiary. This land connection or island archipelago seems to have included the future Greater Antilles that were closely associated with the Nicaraguan region. While this proposal may, of course, not be correct in detail, it is important to appreciate the key point: The very vertebrate groups that were abundant and successful during the Cretaceous (the so-called "Age of Reptiles") have been evolutionarily hot in Central America. In contrast, mammals, which are primarily creatures of the Tertiary (the so-called "Age of Mammals"), have not made much evolutionary headway in Central America. Hence, some material now in Central America and the Antilles appears to have been available to amphibians and reptiles during the Cretaceous before mammals became successful.

The evidence of interchange between North and South America at approximately the upper Cretaceous-Paleocene boundary argues that islands were present in the gap between North and South America at that time. Before that time this material had to be further west, out in the Pacific. But was it exposed and available as habitat earlier in the Cretaceous? Quite possibly, in view of the relatives of cristatellus-bimaculatus anoles in western Mexico and roquet anoles on Malpelo off Columbia, the Pacific distribution of the relatives of the xantusid lizard of Cuba, and the extensive development of endemism in the Central American/Antillean province that Savage has emphasized in the quotations above.

The suture of the southeastern corner of the Caribbean Plate also seems to have taken place in the Pacific, because the Central Venezuelan Basin Fault separates smooth from rough B" material. This material that characterizes the Caribbean sea floor [53] was produced by an offridge episode of volcanism, called the B" event by Burke [52], and the flood basalt event by Donnelly [92], about 80 Ma. It appears to mark the initiation of the Galapagos hot spot [97] when the Caribbean ocean floor was in the Pacific. Given that both sides of the CVBF have B" material, both must have originated in the Pacific, even though they are not identical. The smooth B" sea floor may have been more northern, closer to the Nicaraguan block. The rough B" floor, with islands containing the roquet group of anoles, may have come more from the south, perhaps with affinity to the Cretaceous island arc now part of the Western Cordillera of Columbia and Ecuador [30, 182].

The existence of a Cretaceous Archipelago in the Pacific near South America as a theater where small reptiles evolved and diversified might also be relevant to two of the classic mysteries of herpetology: the occurrence of *Tropidurus* and iguanine lizards on the Galapagos [394] and the iguanine *Brachylophus* of Fiji [120]. Such an archipelago would be well positioned to contribute fauna to both the Galapagos and the Fiji's, which would have been quite close by at the time.

3.4.3 Could this be a big mistake?

A potential value of lineage relationships based on molecular data might be in providing dates for geologic events. The issues are whether the biochemical distance between two populations on separate islands increases as a linear function of time since separation, and if so, what the speed of differentiation is. The mathematical formula for the albumin immunogenetic distance (AID) has been formulated according to population-genetic theory in hopes that this distance does increase linearly with time, thereby providing a molecular clock [389, 213, 388].

Alas, estimates of the clock's speed are either too slow or the ideas expressed in this chapter are fundamentally incorrect. Shochat and Dessauer [331] have suggested that 1.7 units of distance accumulate per 1 million years. Hass and Hedges [134] have suggested a calibration of 1 unit per 0.6 million years, which is equivalent to 1.66 units per 1 million years, thereby agreeing with Shochat and Dessauer. This speed is about twice the rate that seems appropriate for the reconstructions that are proposed in this chapter. Gorman and colleagues [128] noticed this difficulty early on and sided with the molecular clock concluding that the anoles from the gadovi complex of western Mexico somehow resulted from a chance overwater dispersal event from the eastern Caribbean. Hedges and colleagues [142] also side with the molecular clock, and propose that the vertebrates of the West Indies arrived by relatively recent overwater dispersal, and that the absence of an "ancient West Indian [vertebrate] biota" reflects extinctions caused by the same catastrophic bolide impact in the Caribbean region at the close of the Cretaceous that has been implicated in the mass extinctions of the K-T boundary. Clearly, though, this matter is far from settled, and more research is needed.

3.4.4 Historical ecology of Anolis communities

We return now to the ecology of *Anolis* lizards, the subject that is the primary focus of this book. Our findings suggest that the composition of *Anolis* communities in Puerto Rico and the Lesser Antilles results from the combination of competition, habitat bottlenecking, and plate tectonics. Ecological competition prevents cross-colonization by species of the same or very similar body size, but otherwise the strength of competition between anoles is slight. When a geologic fragment splits from a larger unit a very particular set of species is retained on the fragment, those with good "bottle-necking" ability. This faunal fragment of a larger ecological community then develops in isolation, as does a tissue culture extracted from a whole organism. Hence, the Lesser Antilles are not stages on the way toward building up a large community, but are derived extractions from already existing large communities. Conversely, the assembly of large communities, such as those on Cuba and Hispaniola, probably results from combining packages of species when tectonic blocks fuse to form a single island, rather than from the addition of single species, one by one, as studied in the last chapter.

An overall implication of plate tectonics for terrestrial ecology is that relatively fast-acting ecological interactions such as competition and predation are far from sufficient to explain the composition and structure of ecological communities. Instead, ecological communities are fashioned as much by relatively slow geologic processes as by fast species interactions. We have thus come full circle. During the last two decades instances of fast evolution, on an ecological time scale, have been discovered. Now it is also clear that ecological change can itself be very slow, on an evolutionary time scale. Ricklefs and Schluter [276] have emphasized this point as well in an edited volume on historical and geographical perspectives in community ecology. That physical transport processes, such as plate motions in this instance, are as important to community ecology as species interactions is also the essence of "supply-side ecology" [288, 284]. In marine environments, the life cycles of marine populations can only be completed through the intervention of physical transport processes (currents) that convey pelagic larvae to the benthic habitats where they were born. In terrestrial environments, physical transport processes intervene at the community level rather than the population level. Whether a terrestrial community is ecologically saturated depends on whether physical transport processes convey suitable species across the community's boundary. Thus, the actual species composition in a community depends equally on ecological interactions, which determine how much ecological space is available, and on physical transport processes that determine the species pool from which the community's membership is drawn. Unless the transport of eligible species to a community happens to be very high, the community will be less than fully saturated, a point to which we return in the next chapter's discussion.

Finally, we might wonder why the eastern Caribbean has been a theater for the evolution of *Anolis*—why has all this happened there? The answer seems to involve a two-part relationship between anoles and birds. It seems that the present-day dominance of anoles as the ground-feeding insectivorous vertebrates in Caribbean communities reflects a competitive superiority of arboreal lizards over birds in the ground-feeding insectivorous niche, provided the temperature is mild and predation is relatively absent. North American birds are continually migrating through the Lesser Antilles and do not establish there, supporting the idea that the lizards have pre-empted the ground-feeding insectivorous niche in the Lesser Antilles. But larger islands and continental regions support both a higher diversity and a higher abundance of birds of prey. These birds of prey can reduce anole abundance and thereby allow the competitively inferior insectivorous birds to occupy the ground-feeding insectivorous niche. The explanation for the reptile-dominated faunas of the Caribbean, which contrasts markedly with the avian-dominated faunas of the Pacific islands, is that the Caribbean Islands are old enough to have accumulated a reptile fauna before birds and mammals diversified. Because the Caribbean Islands remained at warm latitudes, and because the islands are small enough to lead to relatively little predation on lizards, the islands today may show communities that resemble those of the Cretaceous.

4

The food tangle

I grudgingly concede that there is more to the Caribbean than *Anolis*. This chapter is about the other populations that anoles interact with—their predators, parasites, and prey. This chapter fills out the picture of where anoles fit into the entire ecosystem and raises the issue of how to move beyond a single tier in an ecosystem, as represented by the anoles, to understand how, in some sense, an entire ecosystem functions.

4.1 Food webs

Ecologists are uncertain how to describe an entire ecosystem, and a potentially valuable approach relies on what is called a "food web." To construct a web, draw a vertical link from each prey species to each of its predator species. This leads to a graph with species represented as nodes, and with arrows emanating from the bottom where the producers are, connecting up through various levels such as herbivores, primary and secondary, and even higher consumers, culminating with the decomposers. The strengths of the interactions are not represented, nor are interactions other than trophic interactions, such as interference competition, mutualism, and so forth. Also, a web does not represent the concept of scale, either spatial or temporal, nor hierarchy [236]. In a food web any node logically may be connected to any other node; the representation does not allow nodes to be logically "hidden" from others. Finally, the web does not explicitly represent abiotic mechanisms that may drive the system, controlling its primary productivity and/or mediating the recruitment of larvae to the adult stocks of consumers.

Although highly stylized, the very simplicity of a food web has enabled different kinds of communities to be compared using a common language, and some regularities have been detected [67]. As the database of food webs was enlarged to 113 webs [40] and then to 213 webs [68], empirical support for some of the regularities was strengthened. For example, workers have noticed a rarity of closed loops within the webs; short total chain lengths; a constancy in the proportions of species at basal, intermediate, and top levels; and a similar degree of connectedness from the basal to an intermediate level as from an intermediate level to the top level [178, 177, 311].

The suggestion of patterns among food webs in turn called for some explanatory theory. The first theoretical hypotheses to explain these regularities considered systems of differential equations whose interaction matrix corresponded to a food web. The stability properties of a large community were analyzed as a function of the size of the community, the number of interactions among its members, and the arrangement of those interactions (cf. [248]). This work confirmed the earlier theoretical studies of May [214], predicting a negative relation between diversity and stability, and also accorded with some of the empirical regularities of food webs. Further work has incorporated the role of energy flow [397, 399, 398] and nutrient cycling [81, 82].

More recently, Cohen together with colleagues [69, 72, 73, 70, 230, 74] has offered a model, called the "cascade model" to account for the generalizations of food webs. This model is not a causal hypothesis but a descriptive hypothesis to encapsulate the entire data set on food webs, much in the sense that a regression equation encapsulates a large data set. From a simple assumption concerning the probability that a link occurs between two species, many of the principal generalizations about food webs were derived. Still further topological properties of webs have been described by in [361]. For more detail on these general properties of food webs, readers are referred to summaries in [84, 71, 400].

The development of generalizations about food webs, together with an accompanying theory to account for those food webs, is still a young subject, however. Two criticisms are commonly expressed concerning its current state of development. First, the data for the food webs involve disparate standards for what is lumped into a "trophic species," and these typically reflect disparate standards in the empirical effort that underlies the food webs themselves. Some webs are little more than doodles on a napkin intended to synthesize what is known or believed about a system, while others are the result of comprehensive and painstaking surveys. This variation in the quality of the data, and of standards for presenting the data, gives rise to doubts about the reality of the generalizations from food webs. Second, the 213 food webs used for generalizations about food web structure are mostly small webs. In the ECOWeB data bank [68], the mean web size is smaller than 19 species, and more than 95% of the webs contain fewer than 40 species. Yet North America contains, for instance, about 3500 vertebrate species [375], 90,000 insect species [13], and 16,000 flowering plant species [381]. When groups such as fungi, nematodes, and mites are included, simple arithmetic shows that, unless there are somehow thousands of entirely disjoint food webs in North America, most webs contain dozens to hundreds of species. Therefore, the existing data bank consists of webs that are much smaller than most webs in nature, and it is not clear whether the properties of small webs extrapolate to large webs.

To provide a web that is both detailed and large, the following sections offer a food web for St. Martin. Its construction takes place in two passes, the first somewhat coarse and the second more refined. The size of the final web is 44 species, one of the largest presently available. Then the properties of this web will be contrasted with the current food-web generalizations and discussion offered on where to go from here. This chapter closely follows [121].

4.2 St. Martin web—first pass

Data on the relation of anoles to one another come from the experiments about interspecific competition among anoles, as reviewed in Chapter 2. The data on the relationship of anoles to the trophic level below come from experiments in which lizards were removed, and the response of the arthropod fauna was determined. The data on the relation between anoles and birds come from studies in which the avian predators on anoles were identified and in which the magnitude of the predation on anoles has been documented. Data on the relation between anoles and intestinal parasites come from a survey of the gut parasites of eastern Caribbean anoles that has recently been completed. These studies have led to the food web diagramed in Figure 4.1 and presented as a list data structure in Scheme in Table 4.1. The program foodweb.scm on the diskette can be used to compute statistics for a web when presented in the form of Table 4.1.

4.2.1 Anoles and anoles

As reviewed in Chapter 2, the relation of anoles to one another depends on their body size. Therefore the food webs of the northeastern Caribbean can differ with respect to the *Anolis* component because some of the islands have only one species, some have two, and of those with two species, only on St. Martin is there strong present-day competition.

The web for St. Martin therefore includes arrows coming from a common pool of prey and leading to the two species there. Notice in Figure 4.1 that lines come both from K (small insects on the forest floor) and from a (juvenile spiders) and terminate both in D (*Anolis gingivinus*) and E (*A. pogus*). Also, only D accesses canopy insects e and only E accesses tiny forest-floor insects h. The line from D to E indicates the rare predation by *A. gingivinus* on *A. pogus*.

4.2.2 Anoles and their prey

The first indication of a strong and specific interaction between arthropods and anoles came from Schoener and Toft's discovery [322] of a negative correlation between the abundance of anoles and the abundance of spiders on small islands in the Bahamas. Experimental confirmation of a similar relation was provided by Pacala and Roughgarden [239] on St. Eustatius in the northeastern Caribbean. Schoener and Spiller [321, 340, 342, 341] then performed experiments in the Bahamas that enlarged on the Pacala-Roughgarden protocol and provided more extensive controls. And very recently, Dial [86] has carried out experiments removing lizards from the canopy of rain forest in the Luquillo National Forest of Puerto Rico to see the impact on insects and spiders. The studies show the following main features:

1. Removal of anoles leads to a 10- to 30-fold increase in the number of spiders in the woods where the anoles live. In St. Eustatius the spider webs in the woods approximately 2 m above the ground are especially conspicuous in the experimental treatments.

2. Removal of lizards also leads to an increase of about a factor of 2 in the arthropods on the forest floor. This increase is entirely attributed to an increase of dipterans, which are feeding primarily on fungi in the leaf litter.

3. However, in St. Eustatius no effect of lizard removal was observed on the abundance of leaf-eating insects collected from the vegetation with sweep nets, nor on the amount of leaf damage on the plants in the enclosures from which lizards had been removed. Increased herbivory was observed after lizard removal in the Bahamas and in Puerto Rico. The difference appears to reflect whether the experiments were conducted in habitat where the lizards were feeding primarily on the ground or in the canopy. Thus, on St. Eustatius, and probably on St. Martin as well, the population-dynamic effect of anole predation is primarily on insects that are feeding on the fungi

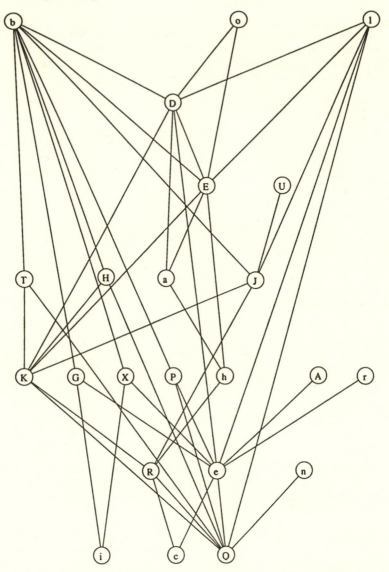

Fig. 4.1 First pass at the St. Martin food web. Each species is represented by a circle with a letter in it. A line connecting two species means that the higher species consumes the lower. The letter assignments and node locations in this figure anticipate those of Figure 4.2. Species coding: A. Adult spider, D. *Anolis gingivinus*, E. *Anolis pogus*, G. Bananaquit H. Bullfinch, J. Big floor insects, K. Small floor insects, P. Elaenia, Q. Fruit and seeds, R. Fungi, T. Grassquit, U. Gray Kingbird, X. Hummingbirds; a. Juvenile spider, b. Kestrel, c. Leaves, e. Canopy insects, h. Tiny floor insects, i. Nectar and floral, l. Pearly-Eyed Thrasher, n. Scaly-Breasted Thrasher, o. Nematodes, r. Yellow Warbler.

Table 4.1 St. Martin food web—first pass.

```
( kestrel (elaenia bullfinch grassquit hummingbirds
          bananaquit big-floor-insects
          anolis-gingivinus anolis-pogus)
  pearly-eyed-thrasher (anolis-pogus anolis-gingivinus
          fruit&seeds canopy-insects big-floor-insects)
  scaly-breasted-thrasher (fruit&seeds)
  gray-kingbird (big-floor-insects)
  elaenia (fruit&seeds canopy-insects)
  bullfinch (fruit&seeds small-floor-insects)
  grassquit (fruit&seeds small-floor-insects)
  hummingbirds (nectar&floral canopy-insects)
  bananaquit (nectar&floral canopy-insects)
  yellow-warbler (canopy-insects)
  anolis-gingivinus (small-floor-insects canopy-insects
                     juvenile-spider anolis-pogus)
  anolis-pogus (tiny-floor-insects small-floor-insects
                juvenile-spider)
  nematodes (anolis-gingivinus anolis-pogus)
  canopy-insects (fruit&seeds leaves)
  big-floor-insects (small-floor-insects fungi)
  small-floor-insects (fungi fruit&seeds)
  tiny-floor-insects (fungi)
  adult-spider (canopy-insects)
  juvenile-spider (tiny-floor-insects)
  fungi (fruit&seeds leaves)
  fruit&seeds ()
  nectar&floral ()
  leaves ()
)
```

in the decomposer subcommunity, and not on insects that are feeding directly on the primary producers.

4. Furthermore, a small indirect effect was discovered. The enclosures with lizards removed, and in which there were more spiders and insects on the forest floor, had about 10% *fewer* insects at middle heights in the forest. These middle heights are where the increase in spider webs is particularly conspicuous. Thus, lizard removal apparently causes an increase in spider abundance that, in turn, causes a lowering of insect abundance in the understory.

These points are reflected so far as possible in Figure 4.1. D (*Anolis gingivinus*) and A (adult spiders) compete by virtue of eating e (the canopy insects). Also, the anoles (D and E) eat the juvenile spiders (a). On removal of lizards these juvenile spiders grow into the abundant adult spiders (A) that, in turn, depress the abundance of the canopy insects (e). The web does not have any way of representing the fact that the juvenile spiders (a) grow to become the adult spiders (A), as though one species is being transformed into another.

4.2.3 Anoles and their predators

In the northern Lesser Antilles the main predators on anoles are avian, and two species are potentially the most important: the American Kestrel (*Falco sparverius*) and the Pearly-Eyed Thrasher (*Margarops fuscatus*) [2]. The predation pressure exerted by these species on *Anolis* lizards has been quantified on the islands of Anguilla, St. Martin, and St. Eustatius [217]. Surprisingly, on Anguilla the feeding habits of the Pearly-Eyed Thrasher shift to a frugivorous from a insectivorous diet and exert no predation pressure on anoles there. The body size of Pearly-Eyed Thrashers also appears slightly smaller on Anguilla than on the nearby islands where lizards do form a significant part of its diet. Moreover, a smaller congener of the Pearly-Eyed Thrasher, the Scaly-Breasted Thrasher (*Margarops fuscus*), that is frugivorous on nearby islands, is absent on Anguilla. Thus, on Anguilla the Pearly-Eyed Thrasher has moved into the Scaly-Breasted Thrasher's niche, leaving the kestrel as the only predator on anoles. But the kestrel's abundance is low enough that the probability of an anole being captured by a kestrel is on the order of 10^{-5} per day, so that anoles themselves are effectively the top predators on this island. In contrast, on St. Martin and St. Eustatius, the Pearly-Eyed Thrasher consumes anoles, and its high abundance combined with a high feeding rate, leads to a predation pressure on anoles of the order of 10^{-3} per day.

Figure 4.1, for St. Martin represents both the American Kestrel (b) and the Pearly-Eyed Thrasher (l) as predators on both anoles. These birds consume insects as well, though generally larger ones than the anoles take. Nonetheless, the Pearly-Eyed Thrasher is probably both a predator and competitor with *Anolis gingivinus*. The known habits of the other birds are also included in the figure.

4.2.4 Anoles and their parasites

The intestinal parasites of *Anolis* in the northern Lesser Antilles are a subset of the larger parasite fauna endemic to the anoles of the Greater Antilles [91]. The main pattern that has emerged is that the parasite load in anoles is strongly correlated with indices of how moist the habitat is: More parasites occur in moist than in dry habitats, as also noted in [304]. This pattern is stronger than any relation to island area, island elevation, and number of anole species. Moreover, the parasite load does not correlate with female fecundity, and its statistical distribution with respect to the body size of the hosts does not indicate any effect on host survivorship. It should be added, however, that these findings are special to the northern Lesser Antillean anoles and not necessarily to anoles elsewhere. The anole of Curacao, *Anolis lineatus*, has at least two orders of magnitude more parasites than Lesser Antillean anoles. Still, even in the northern Lesser Antilles, parasites occur for which anoles are the sole host; are the definitive host, with an arthropod as the intermediate host; and are an intermediate host, with an arthropod as the first intermediate host, and a bird as the definitive host. All these types of parasite life cycles must be congruent with a piece of the food web involving anoles, for the parasites are, in effect, hitchhiking along the links of the web to reach their definitive hosts. In principle, therefore, the web in which *Anolis* occurs should be associated with auxiliary graphs corresponding to the life cycles of these parasites. Anyway, Figure 4.1 shows *o* as a node representing the nematodes that are the gut parasites typically found in anoles.

4.2.5 The food web as "seen" by Anolis

Figure 4.1 as a whole illustrates the web that exists in the hills of St. Martin where two species of anoles coexist and compete, and where two avian predators are significant. As discussed above, however, variations on this web pertain to sea-level sites on St. Martin, where only one anole occurs; on Anguilla where only one anole and one predatory bird occurs; and on St. Eustatius where there are two avian predators and two anoles, but where the anoles differ greatly in body size and have little interspecific competition. The web presents a view of the community as "seen" from the vantage of the anoles (much like posters showing the world as seen from Manhattan).

4.3 St. Martin web—second pass

4.3.1 The food web as "seen" by many

A more comprehensive picture of the food web on St. Martin is obtained, first, by expanding many of the categories that were highly lumped in the preliminary web and, second, by attempting to place weights on each link to indicate its strength. Goldwasser [121] expanded the highly aggregated categories such as big-floor insects and so forth and reconstructed the diets of insects and birds based on published information [29, 13, 140, 297, 378, 379, 79] to obtain a web with 44 species illustrated in Figure 4.2. This expanded web resembles the preliminary web in the top and center, although the nematodes are expanded to three species; most of the expansion is in the lower levels of the web. There is variation in the numbers of prey and predators that a species has. Some species eat and/or are eaten by many species, while others interact with only a few species.

The weights associated with each link are detailed in Table 4.2 (parts A–C) in which the full web is presented as a list data structure. The interpretation of a weighting is as the relative frequency of an act of predation. The predation frequencies were assigned by beginning with the Pearly-Eyed Thrasher and accepting at face value that the predation by thrashers is at a rate of 10 *Anolis* eaten per day

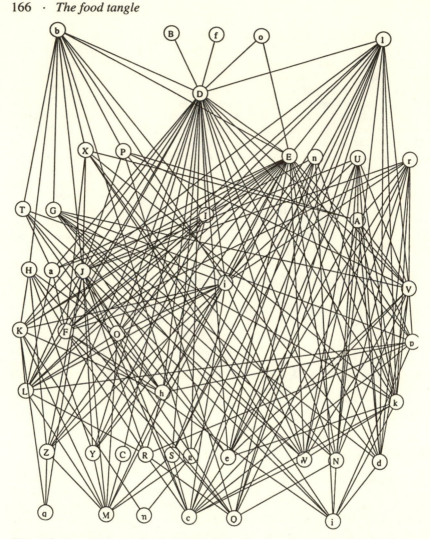

Fig. 4.2 Second pass at the St. Martin food web. Because each species is shown at its minimum height (that is, consuming at least one species in the row immediately below it), the number of rows is one greater than the longest chain in the web. Species codes: A. Adult spider, B. *Allogyptus crenshawi*, C. Annelid, D. *Anolis gingivinus*, E. *Anolis pogus*, F. Ants, G. Bananaquit, H. Bullfinch, I. Centipede, J. Coleoptera adult, K. Coleoptera larva, L. Collembola M. Detritus N. Diptera adult O. Diptera larva P. Elaenia Q. Fruit & seeds, R. Fungi, S. Gastropoda, T. Grassquit, U. Gray Kingbird, V. Hemiptera, W. Homoptera, X. Hummingbirds, Y. Isopoda, Z. Isoptera; a. Juvenile spider, b. Kestrel, c. Leaves, d. Lepidoptera adult, e. Lepidoptera larva, f. *Mesocoelium* sp., g. Millipede, h. Mites, i. Nectar & floral, j. Orthoptera, k. Other hymenoptera, l. Pearly-Eyed Thrasher, m. Roots, n. Scaly-Breasted Thrasher, o. *Thelandros cubensis*, p. Thysanoptera, q. Wood, r. Yellow Warbler.

Table 4.2 St. Martin food web — second pass.

```
( kestrel ((elaenia .003) (bullfinch .003) (grassquit .003)
           (hummingbirds .003) (bananaquit .003)
           (coleoptera-adult .0040) (orthoptera .0143)
           (centipede .0003) (millipede .0004)
           (Anolis-gingivinus .08) (Anolis-pogus .02))
  scaly-breasted-thrasher ((fruit&seeds 95.75)
           (nectar&floral .50) (leaves .25) (centipede .25)
           (other-hymenoptera .50) (gastropoda 2.50)
           (orthoptera .25))
  pearly-eyed-thrasher ((Anolis-pogus 5) (Anolis-gingivinus 5)
           (fruit&seeds 125) (leaves 1) (diptera-adult 1)
           (other-hymenoptera 6) (coleoptera-adult 3) (ants 8)
           (orthoptera 4) (centipede 1.947) (millipede 2.782)
           (lepidoptera-adult .278) (lepidoptera-larva 30.459)
           (gastropoda 5.285) (hemiptera 1.251))
  gray-kingbird ((fruit&seeds 28) (coleoptera-adult 28)
           (other-hymenoptera 21.5) (orthoptera 4)
           (homoptera 3) (hemiptera 6) (diptera-adult .5)
           (lepidoptera-adult 2) (lepidoptera-larva 5)
           (adult-spider 1.5))
  elaenia ((fruit&seeds 86) (adult-spider 6.3)
           (lepidoptera-adult .5) (coleoptera-adult 4.8)
           (hemiptera 2.4))
  yellow-warbler ((orthoptera 7.21) (homoptera 24.84)
           (hemiptera 3.03) (coleoptera-adult 14.62)
           (diptera-adult 8.16) (ants 0.58)
           (lepidoptera-adult 5.98) (lepidoptera-larva 5.97)
           (other-hymenoptera 8.72) (adult-spiders 18.07)
           (fruit&seeds 0.7))
  bullfinch ((fruit&seeds 90) (ants .269)
           (coleoptera-larva .224) (isopoda 2.919)
           (diptera-larva 2.02) (isoptera 4.565))
  grassquit ((fruit&seeds 87.4) (juvenile-spider 1.2)
           (orthoptera 1.2) (homoptera 3.0)
           (lepidoptera-larva 7.2))
  hummingbirds ((nectar&floral 90) (adult-spider 1.13)
           (homoptera 1.48) (ants .37) (coleoptera-adult .08)
           (diptera-adult .71) (other-hymenoptera 6.46))
```

Table 4.2 St. Martin food web—second pass (*continued*).

```
bananaquit ((nectar&floral 90) (diptera-adult 2.98)
        (lepidoptera-adult .042) (lepidoptera-larva 4.60)
        (other-hymenoptera .42) (thysanoptera 1.47)
        (hemiptera .189) (homoptera .29))
thelandros-cubensis ((Anolis-gingivinus 0.095)
        (Anolis-pogus 0.080))
mesocoelium-sp ((Anolis gingivinus .606))
allogyptus-crenshawi ((Anolis gingivinus 0.005))
anolis-gingivinus ((annelid 8) (gastropoda 8) (mites 2)
        (juvenile-spider 10) (centipede 2) (millipede 2)
        (isopoda 4) (coleoptera-adult 148)
        (coleoptera-larva 2) (collembola 2)
         (diptera-adult 56) (ants 852)
        (diptera-larva 20) (hemiptera 2) (homoptera 144)
        (isoptera 54) (other-hymenoptera 212)
        (lepidoptera-adult 36) (lepidoptera-larva 958)
        (orthoptera 12) (thysanoptera 2) (Anolis-pogus 1))
anolis-pogus ((gastropoda 1) (mites 7) (juvenile-spider 33)
        (centipede 1) (isopoda 35) (coleoptera-adult 70)
        (coleoptera-larva 1) (ants 955) (collembola 10)
        (diptera-adult 201) (diptera-larva 7) (hemiptera 2)
        (homoptera 168) (other-hymenoptera 58)
        (lepidoptera-adult 32) (lepidoptera-larva 118)
        (isoptera 7) (orthoptera 11) (thysanoptera 12))
collembola ((fungi 1200) (detritus 1200))
orthoptera ((leaves 180) (detritus 60) (ants 0.20)
        (coleoptera-larva 0.18) (diptera-larva 1.54)
        (isopoda 2.24) (isoptera 3.52) (mites 5.18)
        (collembola 6.92) (homoptera 0.16))
isoptera ((wood 610) (detritus 610))
hemiptera ((leaves 18) (diptera-adult 5.47)
        (thysanoptera 2.70)  (homoptera .54)
        (lepidoptera-adult .08) (lepidoptera-larva 8.44)
        (other-hymenoptera .77))
homoptera ((leaves 56))
thysanoptera ((nectar&floral 70) (leaves 70) (fungi 70)
        (collembola 40) (mites 30))
```

Table 4.2 St. Martin food web—second pass (*continued*).

```
coleoptera-adult ((detritus 8) (fungi 8) (leaves 8)
        (nectar&floral 8) (wood 8)
        (collembola 11.30) (mites 8.47)
        (coleoptera-larva .28) (isopoda 3.67)
        (isoptera 5.74) (diptera-larva 2.54))
coleoptera-larva ((roots 15) (detritus 15) (wood 15)
        (collembola 6.42) (mites 8.57))
ants ((collembola 10.28) (mites 7.71) (fungi 18)
        (nectar&floral 18) (fruit&seeds 18))
other-hymenoptera ((nectar&floral 20) (leaves 20)
        (lepidoptera-larva 37.63) (homoptera 2.40))
lepidoptera-adult ((nectar&floral 8))
lepidoptera-larva ((leaves 476))
diptera-adult ((nectar&floral 284) (fruit&seeds 284))
diptera-larva ((leaves 180) (detritus 180)
        (collembola 102.85) (mites 77.14))
adult-spider ((diptera-adult 28.69)
        (other-hymenoptera 4.04) (thysanoptera 14.14)
        (lepidoptera-adult .40) (lepidoptera-larva 44.24)
        (hemiptera 1.82) (homoptera 2.83))
juvenile-spider ((ants 1.62) (collembola 53.96)
        (mites 40.47))
annelid ((detritus 320))
gastropoda ((leaves 76) (detritus 76))
mites    ((detritus 900) (fungi 900))
centipede ((collembola 19.29) (mites 14.47) (ants .58)
        (coleoptera-larva .48)
        (diptera-larva 4.34) (isopoda 6.27)
        (isoptera 9.81) (juvenile-spider .77))
millipede ((detritus 40) (roots 40))
isopoda ((detritus 780))
fungi    ((fruit&seeds 55) (leaves 55))
fruit&seeds ()
nectar&floral ()
leaves ()
wood ()
roots ()
detritus () )
```

per hectare [217]. The other weightings are scaled relative to this value and take approximate account of abundances and diets. The reason for using predation rates as weights rather than caloric or nutrient content is that this weighting indicates the likelihood that a link will be detected. The weighted web can then be subsampled to simulate food-web studies based on limited field study, so that the possibility of a systematic bias in the ECOWeB database can be investigated [122].

4.3.2 Qualitative properties

Relative to the ECOWeB database of 213 food webs assembled by Cohen [68], the St. Martin web is distinctive in the following respects:

1. St. Martin has more intermediate species, about 70% compared with the 50% typical of ECOWeB.

2. The St. Martin species have more than twice as many links per species, about 5, than ECOWeB, with about 2.

3. The St. Martin web has a maximum chain length of 8; there are 20 chains with this length, although all of them rely on the infrequent link of *Anolis gingivinus* consuming *A. pogus*. The mean chain length is 4.8. These statistics contrast with ECOWeB, in which the average maximum chain length is 3.7, and the average mean chain length is 2.6.

4. The omnivory, calculated as the proportion of species that consumed prey from more than one trophic level, is also higher in the St. Martin web, with 60%, than in ECOWeB, with 25%.

5. Over 80% of the webs in ECOWeB have topological properties called in-tervality, chordality, and an absence of "holes." The St. Martin web is not interval, not chordal, and has holes. Intervality and chordality refer to whether the web is topologically uni-dimensional, and the property of "no holes" refers to whether the species are in a sense tightly packed. All of these feature prominent in the small webs of the ECOWeB database are absent in St. Martin.

Interestingly, the other large food webs that have been put forth recently also differ in much these same ways from the ECOWeB data bank [253, 131, 210]. But it is not size alone that accounts for the differences between the St. Martin web and ECOWeB [121], because the differences cannot be "explained" either by a linear regression of the web statistics against web size, nor by the "cascade model" [72, 73, 70, 230, 74] whose web statistics vary nonlinearly with web size. Thus, the St. Martin web suggests two conclusions. First, existing descriptive generalizations based on the ECOWeB collection cannot be extrapolated, either via linear regression or via the cascade model, to the larger and more complete webs now appearing in the literature. Second, because the St. Martin web fits no current model, it is not clear whether this situation should be ascribed to shortcomings of the current models, to shortcomings of the webs in ECOWeB, or to peculiarities of the St. Martin web itself. This point will be taken up further in the Discussion.

4.3.3 Quantitative properties

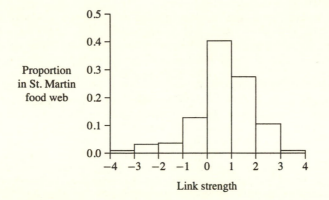

Fig. 4.3 Distribution of link strengths ($\log_{10}$ of frequency per hectare per day) in St. Martin food web.

Figure 4.3 illustrates the distribution of link strengths in the St. Martin web. The horizontal axis is the $\log_{10}$ of the predation frequencies appearing in Table 4.2 (pats A–C). There is no evidence that the link strengths are bimodal, as might be thought if most predators are specialists on just a few types of prey, with only incidental consumption of most other types of prey found in the diets. Instead, most species seem to be rather generalized predators when scored with respect to the taxa of their prey. For example, the anoles overlap almost completely with respect to prey taxa, but prey size and where the prey are actually found are very important in structuring the interaction between anoles. Yet neither prey size nor spatial distribution is representable in a food web diagram.

4.4 Puerto Rico

El Verde web

El Verde, a site in the Luquillo National Forest in Puerto Rico, has supported a large study that has yielded a food web for an island tropical rain forest [372].

The El Verde web consists of 156 nodes that are thought to encompass over 2056 biological species of consumers inhabiting the forest. The distinctive features of the web as contrasted with continental forests under similar climatic conditions are reported as an absence of large herbivores and predators, a generally depauperate flora and fauna, and a superabundance of frogs and lizards. Four food-web patterns are reported: (1) the web contains many closed loops, brought about in large part by cross-predation, whereby generalized consumers eat the juveniles of all species, including their own species and species that compete and prey upon them; (2) the common occurrence of invertebrate predators on vertebrates, made possible because the vertebrates like frogs and lizards are small and within reach of spiders and other invertebrate predators and parasitoids; (3) the widespread presence of omnivory reflecting the generalist characteristics of the consumers when only their diets are taken into account; and (4) a compartmentalization of the web into two relatively unconnected subwebs representing the nocturnal and diurnal components of the ecosystem.

The El Verde web was presented as a large matrix. Its properties were not quantitatively analyzed, and, instead, the matrix is offered to other workers to pursue the analysis. It is possible that the foodweb.scm program could be applied to this data set when the rows of the matrix are unpacked as lists similar to the tables in this chapter, although this has not been attempted.

Workers in the El Verde study emphasize a "nonequilibrium" view of the tropical rain forest, reflecting a focus on the dynamics of the trees. Hurricanes recur in the Caribbean on, say, a 100-year time interval. These hurricanes cause substantial tree fall, which opens up new space and initiates a successional regeneration of the forest. The overall dynamics of the forest impacted recurrently by hurricane disturbance is similar in principle to the dynamics of chaparral and forests recurrently impacted by fire, which also initiate successional sequences. Some care must be used with the adjective of nonequilibrium, however, because whether a system (or subsystem) is usually at equilibrium depends on the ratio of the recurrence time for the disturbance to the time for regeneration. It is possible that the Luquillo trees no sooner complete their regenerative successional sequence than a new hurricane arrives, say every 50 to 100 years or so. In contrast, we have seen that the anoles can come to equilibrium relative to their food supplies in 1 to 2 years. So most of the time the anoles are at an equilibrium that "tracks," or keeps pace with, the slower regeneration time scale of the forest. And the Luquillo forest itself, as a biome, keeps pace with the geological time scale associated with changes in climate and sea level.

Supply-side ecology in the canopy

Lizard removals in the rain forest canopy of the Luquillo National Forest of Puerto Rico [86] revealed what I believe is a surprising fact, that the anoles of the canopy in large part are eating dipterans and other small insects that are blown from the forest floor. The canopies of trees adjacent to, and downwind of, a forest gap have more insects blown into them than trees not adjacent to a gap, or on the upwind side of a gap. As a result, these trees have more lizards in them. Hence, the interaction between lizards and their insect prey may not be a binary interaction. The abundance of anoles depends on the production of insects, but the reciprocal interaction may be weak to nonexistent. The insects blown from the forest floor to the canopy 10 m above may already be lost, denied the opportunity of contributing to the dynamics of the populations from whence they came. Lizard consumption of this "export" may therefore have no effect on insect population dynamics. If so, the lizard-insect interaction may be a unary, not a binary, interaction.

4.5 Discussion

Perhaps the most important generalization to have emerged from over a decade's research on food webs is evidence for a negative relation between diversity and stability [270, 396, 39]. The generalization is that the connectance of a food web decreases as species diversity increases. The connectance of a web is the ratio of the actual number of links to the maximum number of links. A fully connected web has a connectance of 1, while a bunch of noninteracting species has a connectance of 0. It was reported that the bigger a web was, the fewer connections it had within it relative to the maximum number. Each species from all the webs investigated had about the same number of connections. With S species in a web, the number of possible connections goes up as about S^2. The number of actual connections went up as S. So the connectance, which is the ratio of these went down inversely with S.

In 1973, May [214] predicted that a community's stability decreases as its diversity increases, for a given connectance. The only way to achieve a diverse community that is also stable is to have a low connectance. Thus, the food-web generalization appears to validate this prediction.[1]

What would it take to overturn this generalization? To observe no decline of connectance with diversity, the number of links per species would have to increase in direct proportion to S, so that the total number of links would scale as S^2, making the connectance independent of S. In fact, the mean web size in the ECOWeB databank is 19 species, and the average number of links per species is about 2, while the St. Martin web is 44 species and has an average number of links per species of about 5. Taken at face value, if one compares St. Martin to ECOWeB, the number of links per species has about doubled, and the number of species has about doubled too, so that the connectance remains unchanged. Similar doubts about a decline of connectance with diversity have recently been expressed [392, 211].

If this observation continues to be supported in future studies, then it seems that communities are undersaturated to an extent that any relation between diversity and stability is moot. The idea that communities are undersaturated is also increasingly supported in other studies that show local diversity to be positively correlated with regional diversity [78, 275].

Although the generalizations originally drawn about food webs, such as a decline of connectance with diversity, and a universality of certain topological metrics, are not faring well now that better data are being developed, other more process-oriented hypotheses may prove more promising. In the Bahamas, Schoener [311] has pointed out that the maximum chain length seems to depend on island area, with longer chains on larger islands. In a similar vein, Figure 3.3 showed that the species/area relationship among the Virgin Islands is a nested subset pattern. Perhaps the food webs of a sequence of islands of increasing area can be represented as a core trunk, into which successive branches are grafted as the island area increases.

It may be hoped though, that the food-web graph does not become by default the "canonical" representation of an ecological community. Its omission of spatial and temporal scale, and of spatial proximity as the basis for interactions, seems serious. Representations that classify components and processes by scale might be explored instead. At any given time scale, processes occurring an order of magnitude slower than this scale can be regarded as part of the "environment" for that time scale, and processes an order of magnitude faster than this scale might keep up with, or "track," the dynamics of that time scale. Similarly, for a site of any given spatial scale, processes with wider scales may treat the site as an open

---

[1]The number of possible connections is $S(S-1)$ because we don't count a connection of a species to itself. The connecture is $S/(S(S-1))$, which is about $1/S$. A complex strongly connected community is theoretically hard to put together in the first place, and a dearth of strongly connected food webs with high diversity more likely reflects the difficulties of attaining feasibility than stability. A feasible community is an equilibrium solution whose components are all positive. Given positivity, local stability can then be analyzed. Both feasibility itself, and local stability given feasibility, are generally harder to achieve if the connectance is high.

system, while processes with smaller scales are contained within the site. From this perspective, to understand a system one shines a spotlight at a point on the space-scale/time-scale plane and works with the processes and components in a one-order-of-magnitude neighborhood of that point.

The absence of general patterns among food webs suggests we might be more appreciative of natural variation among ecosystems than perhaps we have been, and not push too hard to discover universal generalizations about ecological structure, especially at the community and ecosystem levels. The Hardy-Weinburg law legitimized the variability within species, and made obsolete so-called "typological thinking" whereby a species was identified with an archetypal form and variation was viewed as error [215]. Ecologists have no counterpart of the Hardy-Weinberg law to legitimize the variation across ecosystems, but variation must be faced nonetheless. Although regularities exist, such as niche differences among anoles, such regularities are far from universal generalizations. Indeed, over 60 years of experience from Clements [65] in 1936 to the food-web studies of the last decade suggests that communities and ecosystems are not instances of a universal or ideal ecological system [292]. Because ecology at the community and ecosystem scales is a geographical subject, it is perhaps inappropriate to insist on the discovery of general patterns as a goal. Geological change is a slow chaos, and geography a snapshot of diverse present-day geological states. If geological state strongly influences ecological state, as we think it does, then communities and ecosystems around the world will share little in common as a result. Instead, because the earth is finite and ecological systems are large, nearly exhaustive coverage of the earth can be achieved by understanding ecological systems as particulars, one by one. This particularist outlook does not preclude taking advantage of regularities that do happen to turn up. This outlook does though, emphasize the importance of ecological ease systems.

References

[1] H. Abelson and G. Sussman. *Structure and Interpretation of Computer Programs*. MIT Press, 1985.

[2] S. Adolph and J. Roughgarden. Foraging by passerine birds and *Anolis* lizards on St. Eustatius (Neth. Antilles): implications for interclass competition and predation. *Oecologia*, 56:313–317, 1983.

[3] R. Alatalo, L. Gustafsson, and A. Lundberg. Interspecific competition and niche changes in tits (*Parus* spp.): Evaluation of nonexperimental data. *American Naturalist*, 127:819–834, 1986.

[4] W. W. Anderson. Genetic equilibrium and population growth under density-regulated selection. *American Naturalist*, 105:489–498, 1971.

[5] R. M. Andrews. Growth rate in island and mainland anoline lizards. *Copeia*, 1976:477–482, 1976.

[6] R. M. Andrews. Evolution of life histories: a comparison of *Anolis* lizards from matched island and mainland habitats. *Breviora*, 454:67–72, 1979.

[7] R. M. Andrews. Patterns of growth in reptiles. In C. Gans and F. H. Pough, editors, *Biology of the Reptilia*, pages 273–320. Academic Press, 1982.

[8] R. M. Andrews and T. Asato. Energy utilization of a tropical lizard. *Comparative Biochemistry and Physiology*, 58A:57–62, 1977.

[9] R. M. Andrews and A. S. Rand. Reproductive effort in anoline lizards. *Ecology*, 55:1317–1327, 1974.

[10] R. M. Andrews and A. S. Rand. Seasonal breeding and long-term population fluctuations in the lizard *Anolis limifrons*. In Egbert Leigh, A. S. Rand, and Donald Windsor, editors, *The Ecology of a Tropical Forest: Seasonal Rhythms and Long-term Changes*, pages 405–412. Smithsonian Institution Press, 1982.

[11] R. M. Andrews, A. S. Rand, and S. Guerrero. Seasonal and spatial variation in the annual cycle of a tropical lizard. In A. G. J. Rhodin and K. Miyata, editors, *Advances in Herpetology and Evolutionary Biology*, pages 441–454. Museum of Comparative Zoology, 1983.

[12] J. W. Archie. Methods for coding variable morphological features for numerical taxonomic analysis. *Systematic Zoology*, 34:326–345, 1985.

[13] R. H. Arnett, Jr. *American Insects: A Handbook of the Insects of America North of Mexico*. Van Nostrand Reinhold, New York, 85.

[14] E. N. Arnold. Fossil reptiles from Aldabra atoll, Indian Ocean. *Bull. Brit. Mus. (Nat. Hist.) Zool.*, 29:83–116, 1976.

[15] E. N. Arnold. Indian Ocean giant tortoises: their systematics and island adaptations. *Phil. Trans. R. Soc. B.*, 286:127–145, 1979.

[16] E. N. Arnold. Recently extinct populations from Mauritius and Reunion, Indian Ocean. *J. Zool. Lond.*, 191:33–47, 1980.

[17] R. A. Avery. Observations on habitat utilization by the lizard *Anolis conspersus* on the island of Grand Cayman West Indies. *Amphibia-Reptilia*, 9:417–420, 1988.

[18] R. Baker and H. Genoways. Zoogeography of Antillean bats. In F. B. Gill, editor, *Zoogeography in the Caribbean*, pages 53–97. Spec. Pub. No. 13, Academy of Natural Sciences, 1978.

[19] R. E. Ballinger and D. R. Clark. Energy content of lizard eggs and the measurement of reproductive effort. *Herpetology*, 7:129–132, 1973.

[20] G. A. Bartholomew and V. A. Tucker. Control of changes in body temperature, metabolism, and circulation by the agamid lizard, *Amphibolurus barbatus. Physiological Zoology*, 36:199–218, 1963.

[21] D. J. Beets and H. J. MacGillavry. Outline of the cretaceous and early tertiary history of Curacao, Bonaire, and Aruba. *Guide to Geological Excursions on Curacao, Bonaire, and Aruba, STINAPA Documentation Series No. 2, Caribbean Marine Biological Institute, Curacao*, Series No. 2:1–6, 1977.

[22] D. J. Beets, W. V. Maresch, G. T. Klaver, A. Mottana, R. Bocchio, F. F. Beunk, and H. P. Monen. Magmatic rock series and high-pressure metamorphism as constraints on the tectonic history of the southern Caribbean. *Geological Society of America Memoir*, 162:95–130, 1984.

[23] A. F. Bennett. The energetics of reptilian activity. In C. Gans and F. H. Pough, editors, *Biology of the Reptilia*, pages 155–199. Academic Press, 1982.

[24] A. F. Bennett and W. R. Dawson. Metabolism. In C. Gans and F. H. Pough, editors, *Biology of the Reptilia*, pages 127–223. Academic Press, 1976.

[25] A. F. Bennett, T. T. Gleeson, and G. C. Gorman. Anaerobic metabolism in a lizard *Anolis bonairensis* under natural conditions. *Physiological Zoology*, 54:237–241, 1981.

[26] A. F. Bennett and P. Licht. Anaerobic metabolism during activity in lizards. *Journal of Comparative Physiology*, 81:277–283, 1972.

[27] B. Biju-Duval, A. Mascle, L. Montadert, and J. Wanneson. Seismic investigations in the Columbia, Venezuela and Grenada basins, and on the Barbados Ridge for future ipod drilling. *Geologie en Mijnbouw*, 57:105–116, 1978.

[28] C. Bohlen. Competition for space in the anoles of the island of St. Maarten (Neth. Antilles). unpublished master's thesis, 1983.

[29] D. Borrer, D. DeLong, and C. Triplehorn. *An Introduction to the Study of Insects, Fifth Edition*. Sanders College Publishing, Philadelphia, 1981.

[30] J. Bourgois, J. Toussaint, H. Gonzalez, J. Azema, B. Calle, A. Desmet, L. Murcia, A. Acevedo, E. Parra, and J. Tournon. Geological history of the Cretaceous ophiolitic complexes of northwestern South America (Columbian Andes). *Tectonophysics*, 143:307–327, 1987.

[31] P. Bouysse. Caracteres morphostructuraux et evolution geodynamique de l'arc insulaire des Petites Antilles. *Bureau de Recherches Geologiques et Minieres Bulletin Section IV*, 3/4:185–210, 1979.

[32] P. Bouysse. The Lesser Antilles island arc: Structure and geodynamic evolution. *Initial Reports, Deep Sea Drilling Project (United States Government Printing Office)*, 78A:83–103, 1984.

[33] P. Bouysse, P. Andreieff, M. Richard, J. C. Baubron, A. Mascle, R. C. Maury, and D. Westercamp. Aves Swell and northern Lesser Antilles Ridge: rock-dredging results from ARCANTE 3 cruise. In A. Mascle, editor, *Caribbean Geodynamics*, pages 65–76. Editions Technip, 27 Rue Ginoux, 75737 Paris Cedex 15, 1985.

[34] P. Bouysse, P. Andreieff, and D. Westercamp. Evolution of the Lesser Antilles island arc, new data from the submarine geology. *Transactions of the 9th. Caribbean Geological Conferences (Santo Domingo)*, pages 75–88, 1980.

[35] P. Bouysse, R. Schmidt-Effing, and D. Westercamp. La Desirade island (Lesser Antilles) revisited: Lower Cretaceous radiolarian cherts and arguments against an ophiolitic origin for the basal complex. *Geology*, 11:244–247, 1983.

[36] M. Bowden. Water balance of a dry island. *Geogr. Publ. Dartmouth*, 6:1–89, 1968.

[37] M. S. Boyce. Restitution of r- and k-selection as a model of density-dependent natural selection. *Annual Review of Ecology and Systematics*, 15:427–447, 1984.

[38] B. H. Brattstom. Learning studies in lizards. In N. Greenberg and P. D. MacLean, editors, *Behavior and Neurology of Lizards*, pages 173–181. NIMH, 1978.

[39] F. Briand. Environmental control of food web structure. *Ecology*, 64:253–263, 1983.

[40] F. Briand and J. Cohen. Community food webs have scale-invariant structure. *Nature*, 307:264–266, 1984.

[41] J. C. Briggs. Freshwater fishes and biogeography of Central America and the Antilles. *Systematic Zoology*, 33:428–435, 1984.

[42] G. R. Brooks, Jr. Body temperatures of three lizards from Dominica, West Indies. *Herpetologica*, 24:209–214, 1968.

[43] F. M. Brown. The origins of the West Indian butterfly fauna. In F. B. Gill, editor, *Zoogeography in the Caribbean*, pages 5–30. Academy of Natural Sciences, 1978.

[44] J. H. Brown. Geographical ecology of desert rodents. In M. L. Cody and J. M. Diamond, editors, *Ecology and Evolution of Communities*, pages 315–341. Belknap, Harvard University Press, 1975.

[45] J. S. Brown and T. L. Vincent. A theory for the evolutionary game. *Theoretical Population Biology*, 31:140–166, 1987.

[46] W. L. Brown and E. O. Wilson. Character displacement. *Systematic Zoology*, 5:49–64, 1956.

[47] W. H. Bucher. Geological structure and orogenic history of Venezuela. *Geological Society of America Memoir*, 49, 1952.

[48] D. J. Bullock, E. N. Arnold, and Q. Bloxam. A new endemic gecko (Reptilia: Gekkonidae) from Mauritius. *J. Zool. Lond.*, 206:591–599, 1985.

[49] D. J. Bullock and P. G. H. Evans. The distribution, density and biomass of terrestrial reptiles in Dominica, West Indies. *J. Zool. Lond.*, 222:421–443, 1990.

[50] D. J. Bullock, H. M. Jury, and P. G. H. Evans. Foraging ecology in the lizard *Anolis oculatus* (Iguanidae) from Dominica, West Indies. *J. Zool. Lond.*, 230:19–30, 1993.

[51] G. M. Burghardt. Learning processes in reptiles. In C. Gans and D. W. Tinkle, editors, *Biology of the Reptilia Volume 7*, pages 555–680. Academic Press, 1977.

[52] K. Burke. Tectonic evolution of the Caribbean. *Ann. Rev. Earth Planet. Sci.*, 16:201–230, 1988.

[53] K. Burke, P. J. Fox, and A. Sengor. Buoyant ocean floor and the evolution of the Caribbean. *Journal of Geophysical Research*, 83:3949–3954, 1978.

[54] W. A. Bussing. Geographic distribution of the San Juan ichthyofauna of Central America with remarks on its origin and ecology. In Investigations of the Ichthyofauna of Nicaraguan Lakes, editor, *E. B. Thorson*, page 157. University of Nebraska, 1976.

[55] J. E. Case and T. L. Holcombe. *Geologic-Tectonic Map of the Caribbean Region*. Miscellaneous Investigations Series Map I-1100. U. S. Department of the Interior, U. S. Geological Survey, 1980.

[56] T. J. Case. Character displacement and coevolution in some Cnemidophorus lizards. *Fortschritte der Zoologie*, 25:235–282, 1979.

[57] T. J. Case. Invasion resistance arises in strongly interacting species-rich model competition communities. *Proceedings of the National Academy of Sciences (USA)*, 87:9610–9614, 1990.

[58] T. J. Case. Invasion resistance, species build-up and community collapse in metapopulations models with interspecies competition. *Biol. J. Linn. Soc.*, 42:239–266, 1991.

[59] T. J. Case and M. L. Cody. Testing theories of island biogeography. *American Scientist*, 75:402–411, 1987.

[60] E. L. Charnov. Optimal foraging: the marginal value theorem. *Theoretical Population Biology*, 9:129–136, 1976.

[61] N. Chiariello and J. Roughgarden. Storage allocation in seasonal races of an annual plant: optimal versus actual allocation. *Ecology*, 65:1290–1301, 1984.

[62] D. L. Clark and J. C. Gillingham. Sleep-site fidelity in two Puerto Rican West Indies lizards. *Animal Behavior*, 39:1138–1148, 1990.

[63] B. Clarke. Density-dependent selection. *American Naturalist*, 106:1–13, 1972.

[64] H. T. Claussen and B. Mofshin. The pineal eye of the lizard (*Anolis carolinensis*), a photoreceptor as revealed by oxygen consumption studies. *Journal of Cellular and Comparative Physiology*, 14:29–41, 1939.

[65] F. E. Clements. Nature and structure of the climax. *J. Ecology*, 24:252–284, 1936.

[66] M. L. Cody. *Competition and the Structure of Bird Communities*. Princeton University Press, 1974.

[67] J. E. Cohen. *Food Webs and Niche Space*. Princeton University Press, 1978.

[68] J. E. Cohen. Ecologists' co-operative web bank version 1.0. Machine-readable data base of food webs, Rockefeller University, New York, 1989.

[69] J. E. Cohen. Food webs and community structure. In J. Roughgarden, R. M. May, and S. Levin, editors, *Perspectives in Theoretical Ecology*, pages 181–202. Princeton University Press, 1989.

[70] J. E. Cohen, F. Briand, and C. M. Newman. A stochastic theory of community food webs. III. predicted and observed length of food chains. *Proceedings of the Royal Society of London B, Biological Sciences*, 228:317–353, 1986.

[71] J. E. Cohen, F. Briand, and C. M. Newman. *Community Food Webs: Data and Theory*. Springer-Verlag, Berlin, 1990.

[72] J. E. Cohen and C. M. Newman. A stochastic theory of community food webs. I. models and aggregated data. *Proceedings of the Royal Society of London B, Biological Sciences*, 224:421–448, 1985.

[73] J. E. Cohen, C. M. Newman, and F. Briand. A stochastic theory of community food webs. II. individual webs. *Proceedings of the Royal Society of London B, Biological Sciences*, 224:449–461, 1985.

[74] J. E. Cohen and Z. J. Palka. A stochastic theory of community food webs. V. intervality and triangulation in the trophic niche overlap graph. *American Naturalist*, 135:435–463, 1990.

[75] B. B. Collette. Correlations between ecology and morphology in anoline lizards from Havana, Cuba and southern Florida. *Bulletin of the Museum of Comparative Zoology*, 125:137–162, 1961.

[76] E. D. Cope. On the contents of a bone cave in the island of Anguilla (West Indies). *Smithson. Cont. Knowl*, 489:1–30, 1883.

[77] D. Corke. Reptile conservation on the Maria Islands (St. Lucia, West Indies). *Biological Conservation*, 40:263–270, 1987.

[78] H. V. Cornell. Species assemblages of cypinid gall wasps are not saturated. *American Naturalist*, 126:565–569, 1985.

[79] S. T. Danforth. The birds of Guadeloupe and adjacent islands. *Journal of Agriculture of the University of Puerto Rico*, 2:39–46, 1939.

[80] J. Davis. Growth and size of the western fence lizard (*Sceloporus occidentalis*). *Copeia*, 1967:721–731, 1967.

[81] D. DeAngelis. Energy flow, nutrient cycling, and ecosystem resilience. *Ecology*, 61:764–771, 1980.

[82] D. DeAngelis. *Dynamics of Nutrient Cycling and Food Webs*. Chapman and Hall, 1992.

[83] D. DeAngelis and L. J. Gross, editors. *Individual-based Models and Approaches in Ecology*. Chapman and Hall, 1992.

[84] D. DeAngelis, W. Post, and G. Sugihara, editors. *Current Trends in Food Web Theory*. Oak Ridge National Laboratory, 1983.

[85] G. Dengo and J. E. Case, editors. *The Caribbean Region*, volume H of *The Decade of North American Geology*. Geological Society of America, 3300 Penrose Place, P. O. Box 9140, Boulder Colorado 80301, 1990.

[86] R. Dial. A food web for a tropical rain forest: the canopy view from *Anolis*. Ph. D. Thesis, Stanford University, 1992.

[87] J. M. Diamond. Distributional ecology of New Guinea birds. *Science*, 179:759–769, 1973.

[88] J. M. Diamond. Assembly of species communities. In M. L. Cody and J. M. Diamond, editors, *Ecology and Evolution of Communities*, pages 342–444. Belknap, Harvard University Press, 1975.

[89] J. B. Diebold, P. L. Stoffa, P. Buhl, and M. Truchan. Venezuela basin crustal structure. *Journal of Geophysical Research*, 86:7901–7923, 1981.

[90] L. M. Dill. The escape response of the zebra danio (*Brachydanio rerio*) I. The stimulus for escape. *Animal Behaviour*, 22:711–722, 1974.

[91] A. P. Dobson, S. W. Pacala, J. Roughgarden, E. R. Carper, and E. A. Harris. The parasites of *Anolis* lizards in the northern lesser antilles. I. patterns of distribution and abundance. *Oecologia*, 91:110–117, 1992.

[92] T. W. Donnelly. Mesozoic and Cenozoic plate evolution of the Caribbean region. In F. G. Stehli and S. D. Webb, editors, *The Great American Biotic Interchange*, pages 89–121. Plenum Press, 1985.

[93] J. Dorel. Seismicity and seismic gaps in the Lesser Antilles arc and earthquake hazards in Guadeloupe. *Geophysical Journal of the Royal Astronomical Society*, 67:679–696, 1981.

[94] J. A. Drake. The mechanics of community assembly and succession. *J. Theoretical Biology*, 147:213–233, 1990.

[95] W. E. Duellman. The Central American herpetofauna: an ecology perspective. *Copeia*, 1966:700–719, 1966.

[96] W. E. Duellman. The South American herpetofauna: a panoramic view. *Mus. Nat. Hist. Univ. Kansas Monogr.*, 7:1–28, 1979.

[97] R. A. Duncan and R. Hargraves. Plate tectonic evolution of the Caribbean region in the mantle reference frame. *Geol. Soc. of America, Memoir*, 162:81–93, 1984.

[98] R. A. Duncan and M. A. Richards. Hotspots, mantle plumes, flood basalts, and true polar wander. *Reviews of Geophysics*, 29:31–50, 1991.

[99] A. E. Dunham. Food availability as a proximate factor influencing individual growth rates in the iguanid lizard *Sceloporus merriami*. *Ecology*, 59:770–778, 1978.

[100] A. E. Dunham. An experimental study of interspecific competition between the iguanid lizards *Sceloporus merriami* and *Urosaurus ornatus*. *Ecological Monographs*, 50:309–330, 1980.

[101] J. W. Durham. Movement of the Caribbean plate and its importance for biogeography in the Caribbean. *Geology*, 13:123–125, 1985.

[102] R. K. Dybvig. *The Scheme Programming Language*. Prentice Hall, 1987.

[103] G. C. Eickwort. Distribution patterns and biology of West Indian sweat bees. In J. K. Liebherr, editor, *Zoogeography of Caribbean Insects*, pages 231–253. Comstock Publishing Associates, Cornell University Press, 1988.

[104] M. Eisenberg and H. Abelson. *Programming in Scheme*. The Scientific Press, Redwood City, California, 1988.

[105] J. M. Emlen. The role of time and energy in food preference. *American Naturalist*, 100:611–617, 1966.

[106] R. Estes. The fossil record and early distribution of lizards. In A. G. J. Rhodin and K. Miyata, editors, *Advances in Herpetology and Evolutionary Biology*, pages 365–398. Harvard University, 1983.

[107] R. Etheridge. Late Pleistocene lizards from Barbuda, British West Indies. *Bull. Fla. State Mus.*, 9:43–75, 1964.

[108] R. Etheridge. Fossil lizards from the Dominican Republic. *Quart. J. Fla. Acad. Sci.*, 28:83–105, 1965.

[109] R. Etheridge. An extinct lizard of the genus *Leiocephalus* from Jamaica. *Quart. J. Fla. Acad. Sci.*, 29:47–59, 1966.

[110] D. S. Falconer. *Introduction to Quantitative Genetics, Second Edition*. Longman, London UK, 1981.

[111] T. Fenchel and F. B. Christiansen. Selection and interspecific competition. In T. Fenchel and F. B. Christiansen, editors, *Measuring Selection in Natural Populations. Lecture Notes in Biomathematics*, volume 19, pages 477–498. Springer-Verlag, 1977.

[112] I. Ferrusquia-Villafranca. Distribution of Cenozoic vertebrate faunas in Middle and North America and problems of migration between North and South America. *Inst. Geol. Univ. Auto. Mexico Biol.*, 101:193–321, 1978.

[113] L. K. Fink, Jr. Bathymetric and geologic studies of the Guadeloupe region, Lesser Antilles arc. *Marine Geology*, 12:267–288, 1972.

[114] H. S. Fitch, R. W. Henderson, and H. Guarisco. Aspects of the ecology of an introduced anole: *Anolis cristatellus* in the Dominican Republic. *Amphibia-Reptilia*, 10:307–320, 1989.

[115] K. Fite and B. Lister. Bifoveal vision in *Anolis* lizards. *Brain Behav. Evol.*, 19:144–145, 1981.

[116] L. J. Fleishman. Motion detection in the presence and absence of background motion in an *Anolis* lizard. *Journal of Comparative Physiology A*, 159:711–720, 1986.

[117] C. R. Gallistel. Animal cognition: the representation of space, time, and number. *Annual Review of Psychology*, 40:155–189, 1989.

[118] P. Garrett and S. J. Gould. Geology of New Providence Iisland, Bahamas. *Geological Society of America Bulletin*, 95:209–220, 1984.

[119] N. Ghosh, S. A. Hall, and J. F. Casey. Seafloor spreading magnetic anomalies in the Venezuelan basin. *Geological Society of America Memoir*, 162:65–80, 1984.

[120] J. R. H. Gibbons. The biogeography and evolution of Pacific island reptiles and amphibians. In G. Grigg, R. Shine, and H. Ehmann, editors, *Biology of Australasian Frogs and Reptiles*, pages 125–142. Royal Zoological Society of New South Wales, 1985.

[121] L. Goldwasser and J. Roughgarden. Construction and analysis of a large Caribbean food web. *Ecology*, 74:1216–1233, 1993.

[122] L. Goldwasser and J. Roughgarden. Sampling effects and the estimation of food web properties. Unpublished manuscript, 1993.

[123] G. C. Gorman. The chromosomes of the reptilia, a cytotaxonomic interpretation. In A. Chiarelli and E. Capanna, editors, *Cytotaxonomy and Vertebrate Evolution*, pages 349–421. Academic Press, 1973.

[124] G. C. Gorman and L. Atkins. The zoogeography of Lesser Antillean *Anolis* lizards — an analysis based upon chromosomes and lactic dehydrogenases. *Bull. Mus. Comp. Zool.*, 138:53–80, 1969.

[125] G. C. Gorman, D. G. Buth, M. Soule, and S. Y. Yang. The relationships of the Puerto Rican *Anolis*: electrophoretic and karyotypic studies. In A. G. J. Rhodin and K. Miyata, editors, *Advances in Herpetology and Evolutionary Biology*, pages 626–642. Museum of Comparative Zoology, Cambridge, Mass., 1983.

[126] G. C. Gorman, D. G. Buth, and J. S. Wyles. *Anolis* lizards of the eastern Caribbean: a case study in evolution. III. A cladistic analysis of albumin immunological data, and the definition of species groups. *Syst. Zool.*, 29:143–158, 1980.

[127] G. C. Gorman and Y. J. Kim. *Anolis* lizards – of the eastern Caribbean: a case study in evolution. II. Genetic relationships and genetic variation of the *bimaculatus* group. *Syst. Zool.*, 25:62–77, 1976.

[128] G. C. Gorman, C. S. Lieb, and R. H. Harwood. The relationships of *Anolis gadovi*: albumin immunological evidence. *Carib. J. Sci.*, 20:145–152, 1984.

[129] D. A. Grimaldi. Relects in the drosophilidae (diptera). In J. K. Liebherr, editor, *Zoogeography of Caribbean Insects*, pages 183–213. Comstock Publishing Associates, Cornell University Press, 1988.

[130] C. Guyer and J. Savage. Cladistic relationships among anoles (Sauria: Iguanidae). *Systematic Zoology*, 35:509–531, 1986.

[131] S. J. Hall and D. Raffaelli. Food-web patterns: Lessons from a species-rich web. *Journal of Animal Ecology*, 60:823–842, 1991.

[132] C. B. Harley. Learning the evolutionarily stable strategy. *Journal of Theoretical Biology*, 89:611–633, 1981.

[133] P. H. Harvey and M. Pagel. *The Comparative Method in Evolutionary Biology*. Oxford University Press, 1991.

[134] C. A. Hass and S. B. Hedges. Albumin evolution in West Indian frogs of the genus *Eleutherodactlus* (Leptodactylidae): Caribbean biogeography and a calibration of the albumin immunological clock. *J. Zool., Lond.*, 225:413–426, 1991.

[135] C. A. Hass and S. B. Hedges. Karyotype of the Cuban lizard *Cricosaura typica* and its implications for Xantusiid phylogeny. *Copeia*, 1992:563–565, 1992.

[136] P. J. Hearty and P. Kindler. New perspectives on Bahamian geology: San Salvador Iisland, Bahamas. *J. Coastal Research*, 9:577–594, 1993.

[137] H. Heatwole, T. H. Lin, E. Villalon, A. Muniz, and A. Matta. Some aspects of the thermal ecology of Puerto Rican Anoline lizards. *J. Herpetol.*, 3:65–77, 1969.

[138] H. Heatwole and F. MacKenzie. Herpetogeography of Puerto Rico. IV. Paleogeography, faunal similarity, and endemism. *Evolution*, 21:429–438, 1967.

[139] M. K. Hecht. Fossil lizards of the West Indian genus *Aristelliger* (Gekkonidae). *Amer. Mus. Novit.*, 1538:1–33, 1951.

[140] D. G. Heckel. Extensions of the theory of evolutionary ecology, and the ecology of *Anolis gingivinus*. Ph.D. thesis, Stanford University, 1980.

[141] D. G. Heckel and J. Roughgarden. A technique for estimating the size of lizard populations. *Ecology*, 60:966–975, 1979.

[142] S. B. Hedges, C. A. Hass, and L. R. Maxson. Caribbean biogeography: Molecular evidence for dispersal in West Indian terrestrial vertebrates. *Proc. Nat. Acad. Sci. (USA)*, 89:1909–1913, 1992.

[143] J. Hertz, A. Krogh, and R. Palmer. *Introduction to the Theory of Neural Computation*. Addison-Wesley, 1991.

[144] P. E. Hertz. Sensitivity to high temperatures in three West Indian grass anoles (Sauria, Iguanidae), with a review of heat sensitivity in the genus *Anolis*. *Comp. Biochem. Physiol.*, 63A:217–222, 1979.

[145] P. E. Hertz. Responses to dehydration in *Anolis* lizards sampled along altitudinal transects. *Copeia*, 1980:440–446, 1980.

[146] S. Hillman, G. C. Gorman, and R. Thomas. Water loss in *Anolis* lizards: evidence for acclimation and intraspecific differences along a habitat gradient. *Comp. Biochem. Physiol.*, 62A:491–494, 1979.

[147] J. Holland. *Adaptation in Natural and Artificial Systems*. University of Michigan Press, Ann Arbor, Michigan, 1975.

[148] C. S. Holling. The functional response of invertebrate predators to prey density. *Memoirs of the Entomological Society of Canada*, 48:1–86, 1966.

[149] R. A. Howard. The vegetation of the Antilles. In A. Graham, editor, *Vegetation and Vegetational History of Northern Latin America*, pages 1–38. Elsevier Scientific Publishing Co., 1973.

[150] R. B. Huey. Phylogeny, history and the comparative method. In M. E. Feder, A. F. Bennett, W. W. Burggren, and R. B. Huey, editors, *New Directions in Ecological Physiology*, pages 76–98. Cambridge University Press, 1987.

[151] R. B. Huey and A. F. Bennett. Phylogenetic studies of coadaptation: preferred temperatures versus optimal performance temperatures of lizards. *Evolution*, 41:1098–1115, 1987.

[152] R. B. Huey and P. E. Hertz. Effects of body size and slope on sprint speed of a lizard *(Stellio (Agama) stellio)*. *Journal of Experimental Biology*, 97:401–409, 1982.

[153] R. B. Huey and T. P. Webster. Thermal biology of a solitary lizard: *Anolis marmoratus* of Guadeloupe, Lesser Antilles. *Ecology*, 56:445–452, 1975.

[154] R. B. Huey and T. P. Webster. Thermal biology *Anolis* lizards in a complex fauna: the *Cristatellus* group on Puerto Rico. *Ecology*, 57:985–994, 1976.

[155] P. W. Hummelinck. Studies on the fauna of Curaçao, Aruba, Bonaire and the Venezuelan islands: No. 2. A survey of the mammals, lizards and mollusks. *Fauna of Curaçao, Bonaire and the Venezuelan Islands*, 1:59–108, 1940.

[156] G. E. Hutchinson. Homage to Santa Rosalia, or why are there so many kinds of animals? *American Naturalist*, 93:145–159, 1959.

[157] N. Hotton III. A survey of adaptive relationships of dentition to diet in North American iguanids. *American Midland Naturalist*, 53:88–114, 1955.

[158] Texas Instruments. *PC Scheme User's Guide and Language Reference Manual*. MIT Press, Cambridge Massachussetts, 1990.

[159] Y. Iwasa and J. Roughgarden. Shoot/root balance of plants: optimal growth of a system with many vegetative organs. *Theoretical Population Biology*, 25:78–105, 1984.

[160] A. C. Janetos and B. J. Cole. Imperfectly optimal animals. *Behavioral Ecology and Sociobiology*, 9:203–209, 1981.

[161] T. A. Jenssen, D. L. Marcellini, C. A Pague, and L. A. Jenssen. Competitive interference between two Puerto Rican lizards, *Anolis cooki* and *Anolis cristatellus*. *Copeia*, 1984:853–861, 1984.

[162] A. C. Kamil. Optimal foraging theory and the psychology of learning. *American Zoologist*, 23:291–302, 1983.

[163] A. C. Kamil and H. L. Roitblat. The ecology of foraging behavior: implications for animal learning and memory. *Annual Review of Psychology*, 36:141–169, 1985.

[164] A. C. Kamil and S. I. Yoerg. Learning and foraging behavior. In P. P. G. Bateson and P. H. Klopfer, editors, *Perspectives in Ethology*, pages 325–364. Plenum Press, New York, 1982.

[165] D. King and J. Roughgarden. Graded allocation between vegetative and reproductive growth for annual plants in growing seasons of random length. *Theoretical Population Biology*, 22:1–16, 1982.

[166] W. King and T. Krakauer. The exotic herpetofauna of southeast Florida. *Quart. J. Fla. Acad. Sciences*, 29:144–154, 1966.

[167] J. F. Kitchell and J. T. Windell. Energy budget for the lizard *Anolis carolinensis*. *Physiological Zoology*, 45:178–188, 1972.

[168] J. Koza. *Genetic Programming: On the Programming of Computers by Means of Natural Selection and Genetics*. The MIT Press, Cambridge, Massachusetts, 1992.

[169] J. Koza. The genetic programming paradigm: Genetically breeding populations of computer programs to solve problems. In Branko Soucek and the IRIS Group, editors, *Dynamic, Genetic and Chaotic Programming*, pages 203–321. John Wiley, New York, 1992.

[170] J. Koza, J. Rice, and J. Roughgarden. Evolution of food foraging strategies for the Caribbean *Anolis* lizard using genetic programming. Adaptive Behavior, in press, 1992.

[171] J. Kozłowski. Optimal energy allocation models — an alternative to the concepts of reproductive effort and cost of reproduction. *Acta Œcologica*, 12:11–33, 1991.

[172] J. Kozłowski and J. Uchmański. Optimal individual growth and reproduction in perennial species with indeterminate growth. *Evolutionary Ecology*, 1:214–230, 1987.

[173] J. R. Krebs. Optimal foraging: decision rules for predators. In J. R. Krebs and N. B. Davies, editors, *Behavioral Ecology: An Evolutionary Approach*, pages 23–63. Blackwell, 1978.

[174] J. R. Krebs and N. B. Davies. *Behavioural Ecology. Second Edition*. Blackwell, Oxford, 1984.

[175] D. Lack. *Darwin's Finches*. Cambridge University Press, 1947.

[176] R. Lande. Natural selection and random genetic drift in phenotypic evolution. *Evolution*, 30:314–334, 1976.

[177] J. H. Lawton. Food webs. In J. M. Cherrett, editor, *Ecological Concepts*, pages 43–78. Blackwell Scientific, Oxford, England, 1989.

[178] J. H. Lawton and P. H. Warren. Static and dynamic explanations for patterns in food webs. *Trends in Ecology and Evolution*, 3:242–245, 1988.

[179] J. D. Lazell, Jr. An *Anolis* (Sauria, Iguanidae) in amber. *Journal of Paleontology*, 39:379–382, 1965.

[180] J. D. Lazell, Jr. The Anoles (Sauria, Iguanidae) of the Lesser Antilles. *Bulletin of the Museum of Comparative Zoology (Harvard University)*, 143:1–115, 1972.

[181] J. D. Lazell, Jr. Biogeography of the herpetofauna of the British Virgin Islands, with description of a new anole (Sauria: Iguanidae). In A. G. J. Rhodin and K. Miyata, editors, *Advances in Herpetology and Evolutionary Biology*, pages 99–117. Museum of Comparative Zoology, Cambridge, Mass., 1983.

[182] M. Lebras, F. Megard, C. Dupuy, and J. Dostal. Geochemistry and tectonic setting of pre-collision Cretaceous and Paleogene volcanic rocks of Ecuador. *Geological Society of America Bulletin*, 99:569–578, 1987.

[183] S. C. Levings and D. M. Windsor. Seasonal and annual variation in litter arthropod populations. In Egbert Leigh, A. S. Rand, and Donald Windsor, editors, *The Ecology of a Tropical Forest: Seasonal Rhythms and Long-term Changes*, pages 355–387. Smithsonian Institution Press, 1982.

[184] J. Lewis, G. Draper, C. Bourdon, C. Bowin, P. Mattson, F. Maurrasse, F. Nagle, and G. Pardo. Geology and tectonic evolution of the northern Caribbean margin. In G. Dengo and J. E. Case, editors, *The Caribbean Region*, volume H of *The Decade of North American Geology*, pages 77–140. Geological Society of America, 3300 Penrose Place, P. O. Box 9140, Boulder Colorado 80301, 1990.

[185] P. Licht. Response of *Anolis* lizards to food supplementation in nature. *Copeia*, 1974:215–221, 1974.

[186] P. Licht and G. C. Gorman. Reproductive and fat cycles in Caribbean *Anolis* lizards. *University of California Publications in Zoology*, 95:1–52, 1970.

[187] C. S. Lieb. *Biochemical and Karyological Systematics of the Mexican Lizards of the* Anolis gadovi *and* A. nebulosus *Species Groups*. University Microfilms, 1981.

[188] J. K. Liebherr. Biogeographic patterns of West Indian *Platynus* Carabid beetles (Coleoptera). In J. K. Liebherr, editor, *Zoogeography of Caribbean Insects*, pages 121–152. Comstock Publishing Associates, Cornell University Press, 1988.

[189] Richard P. Lippmann. An introduction to computing with neural nets. *IEEE ASSP Magazine*, 1987:4–22, 1987.

[190] B. C. Lister. Seasonal niche relationships of rain forest anoles. *Ecology*, 62:1548–1560, 1981.

[191] J. B. Losos. Concordant evolution of locomotor behavior display rate and morphology in *Anolis* lizards. *Animal Behavior*, 39:879–890, 1990.

[192] J. B. Losos. Ecomorphology, performance capability, and scaling of West Indian *Anolis* lizards: an evolutionary analysis. *Ecological Monographs*, 60:369–388, 1990.

[193] J. B. Losos. A phylogenetic analysis of character displacement in Caribbean *Anolis* lizards. *Evolution*, 44:558–569, 1990.

[194] J. B. Losos. A critical comparison of the taxon-cycle and character-displacement models for size evolution of *Anolis* lizards in the Lesser Antilles. *Copeia*, pages 279–288, 1992.

[195] J. B. Losos. The evolution of convergent structure in Caribbean *Anolis* communities. Systematic Biology, in press, 1993.

[196] J. B. Losos, R. M. Andrews, O. J. Sexton, and A. L. Schuler. Behavior, ecology, and locomotor performance of the giant anole, *Anolis frenatus*. *Caribbean J. of Science*, 27:173–179, 1991.

[197] J. B. Losos, J. Marks, and T. W. Schoener. Habitat use and ecological interactions of an introduced and a native species of *Anolis* lizard on Grand Cayman. Oecologia, in press, 1993.

[198] J. B. Losos and B. Sinervo. The effects of morphology and perch diameter on sprint performance of *Anolis* lizards. *Journal of Experimental Biology*, 145:23–30, 1989.

[199] R. H. MacArthur. Some generalized theorems of natural selection. *Proceedings of the National Academy (USA)*, 48:1893–1897, 1962.

[200] R. H. MacArthur and R. Levins. The limiting similarity, convergence and divergence of coexisting species. *American Naturalist*, 101:377–385, 1967.

[201] R. H. MacArthur and E. R. Pianka. On the optimal use of a patchy environment. *American Naturalist*, 100:603–609, 1966.

[202] R. H. MacArthur and E. O. Wilson. *The Theory of Island Biogeography*. Princeton University Press, 1967.

[203] A. Malhotra and R. S. Thorpe. Experimental detection of rapid evolutionary response in natural lizard populations. *Nature*, 353:347–348, 1991.

[204] M. Mangel and C. W. Clark. *Dynamic Modeling in Behavioral Ecology*. Princeton University Press, Princeton N. J., 1988.

[205] P. Mann and K. Burke. Neotectonics of the Caribbean. *Reviews of Geophysics and Space Physics*, 22:309–362, 1984.

[206] P. Mann, K. Burke, and T. Matumoto. Neotectonics of Hispaniola: plate motion, sedimentation, and seismicity at a restraining bend. *Earth and Planetary Science Letters*, 70:311–324, 1984.

[207] D. L. Marcellini, T. A. Jenssen, and C. A. Pague. The distribution of *Anolis cooki*, with comments on its possible future extinction. *Herp. Rev.*, 16:99–102, 1985.

[208] W. V. Maresch. Plate tectonics origin of the Caribbean mountain system of northern South America: Discussion and proposal. *Geological Society of America Bulletin*, 85:669–682, 1974.

[209] L. G. Marshall, R. F. Butler, R. E. Drake, G. H. Curtis, and R. H. Tedford. Calibration of the Great American Interchange. *Science*, 204:272–279, 1979.

[210] N. D. Martinez. Artifacts or attributes? effects of resolution on the Little Rock Lake food web. *Ecological Monographs*, 61:367–392, 1991.

[211] N. D. Martinez. Constant connectance in community food webs. *American Naturalist*, 139:1208–1218, 1992.

[212] A. Mascle and P. Letouzey. *Geological Map of the Caribbean*. Institut Français du Pétrole, Editions Technip 27, rue Ginoux, 75737 Paris Cedex 15, 1990.

[213] L. R. Maxson and A. C. Wilson. Albumin evolution and organismal evolution in tree frogs (Hylidae). *Systematic Zoology*, 24:1–15, 1975.

[214] R. May. *Stability and Complexity in Model Ecosystems*. Princeton University Press, 1973.

[215] E. Mayr. *Animal Species and Evolution*. Belknap Press of Harvard University Press, 1963.

[216] D. A. McFarlane and R. D. E. MacPhee. *Amblyrhiza* and the Quarternary bone caves of Anguilla, British West Indies. *Cave Science*, 16:31–34, 1989.

[217] J. F. McLaughlin and J. Roughgarden. Avian predation on *Anolis* lizards in the northeastern Caribbean: an inter-island contrast. *Ecology*, 70:617–628, 1989.

[218] J. M. McNamara and A. Houston. Optimal foraging and learning. *J. Theor. Biol.*, 117:231–249, 1985.

[219] J. M. McNamara and A. Houston. Memory and efficient use of information. *J. Theor. Biol.*, 125:385–395, 1987.

[220] Z. P. Metcalf. The center of origin theory. *Elisha Micthel Sci. Soc. J.*, 62:149–175, 1947.

[221] D. Miles and A. E. Dunham. Comparative analyses of phylogenetic effects in the life history patterns of iguanid reptiles. *American Naturalist*, 139:848–869, 1992.

[222] D. Miles and A. E. Dunham. Evolution of body size in island *Anolis* and the importance of phylogeny of ecological analyses. Unpublished manuscript, 1992.

[223] T. C. Moermond. Habitat constraints on the behavior, morphology, and community structure of *Anolis* lizards. *Ecology*, 60:152–164, 1979.

[224] T. C. Moermond. The influence of habitat structure on *Anolis* foraging behavior. *Behaviour*, 70:147–167, 1979.

[225] T. C. Moermond. Prey attack behavior of *Anolis* lizards. *Zeitschrift fur Tierpsychologie*, 56:128–136, 1981.

[226] T. C. Moermond. A mechanistic approach to the structure of animal communities: *Anolis* lizards and birds. *American Zoologist*, 26:23–37, 1986.

[227] M. Moulton and S. Pimm. The extent of competition in shaping an introduced avifauna. In J. Diamond and T. Case, editors, *Community Ecology*, pages 80–97. Harper and Row, 1986.

[228] L. D. Mueller and F. J. Ayala. Trade-off between r-selection and K-selection in *Drosophila* populations. *Proc. Nat. Acad. Sci. (USA)*, 78:1303–1305, 1981.

[229] K. H. Naganuma and J. Roughgarden. Optimal body size in Lesser Antillean *Anolis* lizards — a mechanistic approach. *Ecological Monographs*, 60:239–256, 1990.

[230] C. M. Newman and J. E. Cohen. A stochastic theory of community food webs. IV. theory of food chain lengths in large webs. *Proceedings of the Royal Society of London B, Biological Sciences*, 228:354–377, 1986.

[231] S. W. Nichols. Kaleidoscopic biogeography of West Indian Scaritinae (Coleoptera: Carabidae). In J. K. Liebherr, editor, *Zoogeography of Caribbean Insects*, pages 71–120. Comstock Publishing Associates, Cornell University Press, 1988.

[232] H. T. J. Norton. Natural selection and Mendelian variation. *Proceedings of the London Mathematical Society*, 28:1–45, 1928.

[233] J. G. Ollason. Learning to forage—optimally? *Theoretical Population Biology*, 18:44–56, 1980.

[234] S. L. Olson. A paleontological perspective of West Indian birds and mammals. In F. B. Gill, editor, *Zoogeography in the Caribbean*, pages 99–117. Spec. Pub. No. 13, Academy of Natural Sciences, 1978.

[235] S. L. Olson and H. F. James. Fossil birds from the Hawaiian Islands: evidence for wholesale extinction by man before western contact. *Science*, 717:633–635, 1982.

[236] R. V. O'Neill, D. L. DeAngelis, J. B. Waide, and T. F. H. Allen. *A Hierarchical Concept of Ecosystems*. Princeton University Press, 1986.

[237] S. W. Pacala and J. Roughgarden. The evolution of resource partitioning in a multidimensional resource space. *Theoretical Population Biology*, 22:127–145, 1982.

[238] S. W. Pacala and J. Roughgarden. Resource partitioning and interspecific competition in two two-species insular *Anolis* lizard communities. *Science*, 217:444–446, 1982.

[239] S. W. Pacala and J. Roughgarden. Control of arthropod abundance by *Anolis* lizards on St. Eustatius (Neth. Antilles). *Oecologia*, 64:160–162, 1984.

[240] S. W. Pacala and J. Roughgarden. Population experiments with the *Anolis* lizards of St. Maarten and St. Eustatius. *Ecology*, 66:129–141, 1985.

[241] S. W. Pacala, J. D. Rummel, and J. Roughgarden. A technique for enclosing *Anolis* populations under field conditions. *Journal of Herpetology*, 17:94–97, 1983.

[242] G. Packard, C. Tracy, and J. Roth. The physiological ecology of reptilian eggs and embryos, and the evolution of vivipary within the class Reptilia. *Biol. Rev.*, 52:71–105, 1977.

[243] M. Pagel and P. H. Harvey. Recent developments in the analysis of comparative data. *Quart. Rev. Biol.*, 63:413–440, 1988.

[244] D. Paull, E. E. Williams, and W. P. Hall. Lizard karyotypes from the Galapagos Islands: chromosomes in phylogeny and evolution. *Breviora*, 441:1–31, 1976.

[245] M. R. Perfit and E. E. Williams. Geological constraints and biological retrodictions in the evolution of the Caribbean Sea and its islands. In C. A. Woods, editor, *Biogeography of the West Indies*, pages 47–102. Sandhill Crane Press, 1989.

[246] R. H. Peters. *The Ecological Implications of Body Size*. Cambridge University Press, 1983.

[247] E. R. Pianka. *Ecology and Natural History of Desert Lizards*. Princeton University Press, 1986.

[248] S. Pimm. *Food Webs*. Chapman and Hall, 1982.

[249] J. Pindell and J. F. Dewey. Permo-Triassic reconstruction of western Pangea and the evolution of the Gulf of Mexico/Caribbean region. *Tectonics*, 1:179–211, 1982.

[250] J. L. Pindell and S. F. Barrett. Geological evolution of the Caribbean region; a plate-tectonic perspective. In G. Dengo and J. E. Case, editors, *The Caribbean Region*, volume H of *The Decade of North American Geology*, pages 405–432. Geological Society of America, 3300 Penrose Place, P. O. Box 9140, Boulder Colorado 80301, 1990.

[251] B. Pinet, D. Lajat, P. Le Quellec, and Ph. Bouysse. Structure of Aves Ridge and Grenada Basin from multichannel seismic data. In A. Mascle, editor, *Caribbean Geodynamics*, pages 53–64. Editions Technip, 27 Rue Ginoux, 75737 Paris Cedex 15, 1985.

[252] G. O. Poinar, Jr. and D. C. Cannatella. An Upper Eocene frog from the Dominican Republic and its implications for Caribbean biogeography. *Science*, 237:1215–1216, 1987.

[253] G. A. Polis. Complex trophic interactions in deserts: an empirical assessment of food-web theory. *American Naturalist*, 138:123–155, 1991.

[254] F. H. Pough. The advantages of ectothermy for tetrapods. *American Naturalist*, 115:92–112, 1980.

[255] R. Powell, D. D. Smith, J. S. Parmerlee, C. V Taylor, and M. L. Jolley. Range expansion by an introduced anole: *Anolis porcatus* in the dominican republic. *Amphibia-Reptilia*, 11:421–425, 1990.

[256] G. K. Pregill. Late Pleistocene herpetofaunas from Puerto Rico. *Misc. Publ. Univ. Kansas Mus. Nat. Hist.*, 71:1–72, 1981.

[257] G. K. Pregill. Fossil amphibians and reptiles from New Providence, Bahamas. *Smithson. Contrib. Paleobiol.*, 48:8–22, 1982.

[258] G. K. Pregill. An extinct species of *Leiocephalus* from Haiti (Sauria: Iguanidae). *Proc. Biol. Soc. Wash.*, 97:827–833, 1984.

[259] G. K. Pregill. Body size of insular lizards: a pattern of Holocene dwarfism. *Evolution*, 40:997–1008, 1986.

[260] H. R. Pulliam. On the theory of optimal diets. *American Naturalist*, 108:59–74, 1974.

[261] G. H. Pyke. Optimal foraging: a critical review. *Annual Review of Ecology and Systematics*, 15:523–575, 1984.

[262] J. A. Ramos. Zoogeography of the Auchenorrhynchous Homoptera of the Greater Antilles (Hemiptera). In J. K. Liebherr, editor, *Zoogeography of Caribbean Insects*, pages 61–70. Comstock Publishing Associates, Cornell University Press, 1988.

[263] A. S. Rand. Ecological distribution of the anoline lizards of Puerto Rico. *Ecology*, 45:745–752, 1964.

[264] A. S. Rand. Ecological distribution of the anoline lizards around Kingston, Jamaica. *Breviora*, 22:1–18, 1967.

[265] A. S. Rand, G. C. Gorman, and W. M. Rand. Natural history, behavior, and ecology of *Anolis agassizi*. *Smithsonian Contributions to Zoology*, 176:27–38, 1975.

[266] D. P. Reagan. Foraging behavior of *Anolis stratulus* in a Puerto Rican rain forest. *Biotropica*, 18:157–160, 1986.

[267] D. P. Reagan. Congeneric species distribution and abundance in a three-dimensional habitat: the rain forest anoles of Puerto Rico. *Copeia*, 1992:392–403, 1992.

[268] P. J. Regal. Behavioral differences between reptiles and mammals: an analysis of activity and mental capabilities. In N. Greenberg and P. D. MacLean, editors, *Behavior and Neurology of Lizards*, pages 183–202. NIMH, 1978.

[269] D. E. Reichle. Energy and nutrient metabolism of soil and litter invertebrates. In P. Duvigneaud, editor, *Productivity of forest ecosystems*, pages 465–475. UNESCO, 1971.

[270] M. Rejmanek and P. Stary. Connectance in real biotic communities and critical values for stability of model ecosystems. *Nature*, 280:311–313, 1979.

[271] P. R. Renne, J. M. Mattinson, C. W. Hatten, M. Somin, T. C. Onstott, G. Millan, and E. Linares. ^{40}Ar/^{39} and U-Pb evidence for late proterozoic (Grenville-age) continental crust in northcentral Cuba and regional tectonic implications. *Precambrian Research*, 42:325–341, 1989.

[272] P. L. Richardson and D. Walsh. Mapping climatological seasonal variations of surface currents in the tropical Atlantic using ship drifts. *J. Geophys. Res.*, 91:10537–10550, 1986.

[273] A. D. Richman, T. J. Case, and T. Schwaner. Natural and unnatural extinction rates of reptiles on islands. *American Naturalist*, 131:611–630, 1988.

[274] R. E. Ricklefs. Energetics of reproduction in birds. In R. A. Paynter, editor, *Avian Energetics*, pages 152–297. Nuttall Ornithological Club, 1974.

[275] R. E. Ricklefs. Community diversity: relative roles of local and regional processes. *Science*, 235:167–171, 1987.

[276] R. E. Ricklefs and D. Schluter. *Species Diversity in Ecological Communities, Historical and Geographical Perspectives*. Chicago University Press, 1993.

[277] O. Rieppel. Green anole in Dominican amber. *Nature*, 286:486–487, 1980.

[278] L. E. Rogers, R. L. Buschbom, and C. R. Watson. Length-weight relationships of shrub-steppe invertebrates. *Annals of the Entomological Society of America*, 70:51–53, 1976.

[279] B. Rose. Food intake and reproduction in *Anolis acutus*. *Copeia*, 1982:322–330, 1982.

[280] J. Roughgarden. Density-dependent natural selection. *Ecology*, 52:453–468, 1971.

[281] J. Roughgarden. Evolution of niche width. *American Naturalist*, 106:683–718, 1972.

[282] J. Roughgarden. Niche width: biogeographic patterns among *Anolis* lizard populations. *American Naturalist*, 108:429–442, 1974.

[283] J. Roughgarden. *Theory of Population Genetics and Evolutionary Ecology: An Introduction*. Macmillan, 1979.

[284] J. Roughgarden. The structure and assembly of communities. In J. Roughgarden, R. M. May, and S. A. Levin, editors, *Perspectives in Theoretical Ecology*, pages 203–226. Princeton University Press, 1989.

[285] J. Roughgarden. Origin of the eastern Caribbean: data from reptiles and amphibians. In D. K. Larue and G. Draper, editors, *Transactions of the 12th Caribbean Geological Conference, St. Croix, USVI*, pages 10–26. Miami Geological Society, 1990.

[286] J. Roughgarden. The evolution of sex. *American Naturalist*, 138:934–953, 1991.

[287] J. Roughgarden and Eduardo R. Fuentes. The environmental determinants of size in solitary populations of West Indian *Anolis* lizards. *Oikos*, 29:44–51, 1977.

[288] J. Roughgarden, S. Gaines, and S. W. Pacala. Supply side ecology: the role of physical transport processes. In P. Giller and J. Gee, editors, *Organization of Communities: Past and Present*, pages 459–486. Blackwell Scientific Publications, 1987.

[289] J. Roughgarden, D. G. Heckel, and Eduardo R. Fuentes. Coevolutionary theory and the biogeography and community structure of *Anolis*. In R. Huey, E. R. Pianka, and T. W. Schoener, editors, *Lizard Ecology: Studies of a Model Organism*, pages 371–410. Harvard University Press, 1983.

[290] J. Roughgarden and S. W. Pacala. Taxon cycle among *Anolis* lizard populations: review of evidence. In D. Otte and J. Endler, editors, *Speciation and its Consequences*, pages 403–432. Sinauer, 1989.

[291] J. Roughgarden, S. W. Pacala, and J. D. Rummel. Strong present-day competition between the *Anolis* lizard populations of St. Maarten (Neth. Antilles). In B. Shorrocks, editor, *Evolutionary Ecology*, pages 203–220. Blackwell Scientific Publications, 1984.

[292] J. Roughgarden, T. Pennington, and S. Alexander. Dynamics of the rocky intertidal zone with remarks on generalization in ecology. *Philosophical Transactions of the Royal Society (London) Series B*, 343:79–85, 1994.

[293] J. Roughgarden, Warren Porter, and D. G. Heckel. Resource partitioning of space and its relationship to body temperature in *Anolis* lizard populations. *Oecologia*, 50:256–264, 1981.

[294] J. Roughgarden, J. D. Rummel, and S. W. Pacala. Experimental evidence of strong present-day competition between the *Anolis* populations of the Anguilla Bank—a preliminary report. In A. Rhodin and K. Miyata, editors, *Advances in Herpetology and Evolutionary Biology—Essays in Honor of Ernest Williams*, pages 499–506. Museum of Comparative Zoology, 1983.

[295] I. Rouse and L. Allaire. Caribbean. In R. E. Taylor and C. W. Meighan, editors, *Chronologies in New World Archeology*, pages 431–483. Academic Press, 1978.

[296] R. Ruibal, R. Philibosian, and J. Adkins. Reproductive cycle and growth in the lizard *Anolis acutus*. *Copeia*, 1972:509–518, 1972.

[297] J. D. Rummel. Investigations into interspecific competition and evolutionary theory. Ph.D. thesis, Stanford University, 1984.

[298] J. D. Rummel and J. Roughgarden. Effects of reduced perch-height separation on competition between two *Anolis* lizards. *Ecology*, 66:430–444, 1985.

[299] J. D. Rummel and J. Roughgarden. A theory of faunal buildup for competition communities. *Evolution*, 39:1009–1033, 1985.

[300] M. A. Salzburg. *Anolis sagrei* and *Anolis cristatellus* in southern Florida: a case study in interspecific competition. *Ecology*, 65:14–19, 1984.

[301] J. M. Savage. The origins and history of the Central American herpetofauna. *Copeia*, 1966:719–766, 1966.

[302] J. M. Savage. The Isthmian Link and the evolution of neotropical mammals. *Los Angeles County Mus. Contr. Sci.*, 160:1–51, 1974.

[303] J. M. Savage. The enigma of the Central American herpetofauna: dispersals or vicariance? *Ann. Missouri Bot. Gard.*, 69:464–547, 1982.

[304] J. J. Schall. Parasite-mediated competition in *Anolis* lizards. *Oecologia*, 92:58–64, 1992.

[305] D. Schluter and P. Grant. Determinants of morphological patterns in communities of Darwin's finches. *American Naturalist*, 123:175–196, 1984.

[306] T. W. Schoener. Size patterns in West Indian *Anolis* lizards. I. Size and species diversity. *Systematic Zoology*, 18:386–401, 1969.

[307] T. W. Schoener. Size patterns of West Indian *Anolis* lizards. II. Correlations with the sizes of particular sympatric species-displacement and convergence. *American Naturalist*, 104:155–174, 1970.

[308] T. W. Schoener. Theory of feeding strategies. *Annual Review of Ecology and Systematics*, 2:369–404, 1971.

[309] T. W. Schoener. Length-weight regressions in tropical and temperate forest-understory insects. *Annals of the Entomological Society of America*, 73:106–109, 1980.

[310] T. W. Schoener. A brief history of optimal foraging ecology. In A. C. Kamil, J. R. Krebs, and H. R. Pulliam, editors, *Foraging Behavior*, pages 5–67. Plenum Press, New York, 1987.

[311] T. W. Schoener. Food webs from the small to the large. *Ecology*, 70:1559–1589, 1989.

[312] T. W. Schoener and G. C. Gorman. Some niche differences in three Lesser Antillean lizards of the genus *Anolis*. *Ecology*, 49:819–830, 1968.

[313] T. W. Schoener and D. H. Janzen. Notes on environmental determinants of tropical versus temperate insect size patterns. *American Naturalist*, 102:207–224, 1968.

[314] T. W. Schoener and A. Schoener. Structural habitats of West Indian *Anolis* lizards. I. Jamaican lowlands. *Breviora*, 368:1–53, 1971.

[315] T. W. Schoener and A. Schoener. Structural habitats of West Indian *Anolis* lizards. II. Puerto Rican uplands. *Breviora*, 375:1–39, 1971.

[316] T. W. Schoener and A. Schoener. Estimating and interpreting body-size growth in some *Anolis* lizards. *Copeia*, 1978:390–405, 1978.

[317] T. W. Schoener and A. Schoener. The ecological correlates of survival in some Bahamian *Anolis* lizards. *Oikos*, 39:1–16, 1982.

[318] T. W. Schoener and A. Schoener. Intraspecific variation in home-range size in some *Anolis* lizards. *Ecology*, 63:809–823, 1982.

[319] T. W. Schoener and A. Schoener. Distribution of vertebrates on some very small islands. I. occurrence sequences of individual species. *Journal of Animal Ecology*, 52:209–235, 1983.

[320] T. W. Schoener and A. Schoener. Distribution of vertebrates on some very small islands. II. patterns in species number. *Journal of Animal Ecology*, 52:237–262, 1983.

[321] T. W. Schoener and D. Spiller. Effect of lizards on spider populations: manipulative reconstruction of a natural experiment. *Science*, 236:949–952, 1987.

[322] T. W. Schoener and C. Toft. Spider populations: extraordinarily high densities on islands without top predators. *Science*, 219:1353–1355, 1983.

[323] A. Schwartz. Some aspects of the herpetogeography of the West Indies. In F. B. Gill, editor, *Zoogeography in the Caribbean*, pages 31–51. Spec. Pub. No. 13, Academy of Natural Sciences, 1978.

[324] A. Schwartz and R. W. Henderson. *A Guide to the Identification of the Amphibians and Reptiles of the West Indies Exclusive of Hispaniola*. Milwaukee Public Museum, 1985.

[325] A. Schwartz and R. Thomas. A check-list of West Indian amphibians and reptiles. *Carnegie Mus. Nat. Hist. Spec. Publ.*, 1:1–216, 1975.

[326] O. J. Sexton. Population changes in a tropical lizard *Anolis limifrons* on Barro Colorado Island, Panama Canal Zone. *Copeia*, 1967:219–222, 1967.

[327] O. J. Sexton. Reproductive cycles of Panamanian reptiles. In William G. D'Arcy and Mireya D. Correa A., editors, *The Botany and Natural History of Panama: La Botánica e Historia Natural de Panamá*, pages 95–110. Missouri Botanical Garden, 1985.

[328] O. J. Sexton, J. Bauman, and E. Ortleb. Seasonal food habits of *Anolis limifrons*. *Ecology*, 53:182–186, 1972.

[329] O. J. Sexton, E. Ortleg, L. Hathaway, R. Bollinger, and P. Licht. Reproductive cycles of three species of anoline lizards from the Isthmus of Panama. *Ecology*, 52:201–215, 1971.

[330] S. Shafir and J. Roughgarden. Instrumental discrimination conditioning of *Anolis cristatellus* in the field with food as a reward. Carribean J. of Science, in press, 1994.

[331] D. Shochat and H. C. Dessauer. Comparative immunological study of albumins of *Anolis* lizards of the Caribbean Islands. *Comp. Biochem. Physiol.*, 68A:67–73, 1981.

[332] C. Singer. *A Short History of Biology*. Oxford Clarendon Press, 1931.

[333] J. A. Slater. Zoogeography of West Indian Lygaeidae (Hemiptera). In J. K. Liebherr, editor, *Zoogeography of Caribbean Insects*, pages 38–60. Comstock Publishing Associates, Cornell University Press, 1988.

[334] M. Slatkin. On the equilibration of fitnesses by natural selection. *American Naturalist*, 112:845–859, 1978.

[335] M. Slatkin. Ecological character displacement. *Ecology*, 61:163–178, 1980.

[336] N. H. Sleep. Hotspots and mantle plumes: some phenomenology. *Journal of Geophysical Research*, 95(B5):6715–6736, 1990.

[337] J. Smith. *An Introduction to Scheme*. Prentice Hall, 1988.

[338] R. C. Speed. Cenozoic collision of the Lesser Antilles arc and continental South America. *Tectonics*, 4:41–69, 1985.

[339] R. C. Speed, L. C. Gerhard, and E. H. McKee. Ages of deposition, deformation, and intrusion of Cretaceous rocks, eastern St. Croix, Virgin Islands. *Geological Society of America Bulletin I*, 90:629–632, 1979.

[340] D. A. Spiller and T. W. Schoener. An experimental study of the effect of lizards on web-spider communities. *Ecological Monographs*, 58:57–77, 1988.

[341] D. A. Spiller and T. W. Schoener. Lizards reduce food consumption by spiders: mechanisms and consequences. *Oecologia*, 83:150–161, 1990.

[342] D. A. Spiller and T. W. Schoener. A terrestrial field experiment showing the impact of eliminating top predators on foliage damage. *Nature*, 347:469–472, 1990.

[343] J. A. Stamps. Displays and social organization in female *Anolis aeneus*. *Copeia*, 1973:264–272, 1973.

[344] J. A. Stamps. Courtship patterns, estrus periods and reproductive condition in a lizard, *Anolis aeneus*. *Physiol. Behav.*, 14:531–535, 1975.

[345] J. A. Stamps. Egg retention, rainfall, and egg laying in a tropical lizard *Anolis aeneus*. *Copeia*, 1976:759–764, 1976.

[346] J. A. Stamps. Rainfall, moisture, and dry season growth rates in *Anolis aeneus*. *Copeia*, 1977:415–419, 1977.

[347] J. A. Stamps. A field study of the ontogeny of social behavior in the lizard *Anolis aeneus*. *Behavior*, 66:1–31, 1978.

[348] J. A. Stamps. The relationship between ontogenetic habitat shifts, competition and predator avoidance in a juvenile lizard (*Anolis aeneus*). *Behavioral Ecology and Sociobiology*, 12:19–33, 1983.

[349] J. A. Stamps. Territoriality and the defense of predator refuges in juvenile lizards *Anolis aeneus*. *Animal Behavior*, 31:857–870, 1983.

[350] J. A. Stamps. Rank-dependent compromises between growth and predator protection in lizard *Anolis aeneus* dominance hierarchies. *Animal Behavior*, 32:1101–1107, 1985.

[351] J. A. Stamps. Conspecifics as cues to territory quality: a preference of juvenile lizards *Anolis aeneus* for previously used territories. *American Naturalist*, 129:629–642, 1987.

[352] J. A. Stamps. The effect of body size on habitat and territory choice in juvenile lizards. *Herpetologica*, 44:369–376, 1988.

[353] J. A. Stamps and D. Crews. Seasonal changes in reproduction and social behavior in the lizard *Anolis aeneus*. *Copeia*, 1976:467–477, 1976.

[354] J. A. Stamps and P. K. Eason. Relationships between spacing behavior and growth rates: a field study of lizard feeding territories. *Behavioral Ecology and Sociobiology*, 25:99–108, 1989.

[355] J. A. Stamps, S. Tanaka, and V. V. Krishnan. Relationship between selectivity and food abundance in a juvenile lizard *Anolis aeneus*. *Ecology*, 62:1079–1092, 1981.

[356] J. A. Stamps and S. K. Tanaka. The influence of food and water on growth rates in a tropical lizard (*Anolis aeneus*. *Ecology*, 62:33–40, 1981.

[357] D. W. Steadman, G. K. Pregill, and S. L. Olson. Fossil vertebrates from Antigua, Lesser Antilles: evidence for late Holocene human extinctions in the West Indies. *Proc. Natl. Acad. Sci.*, 81:4448–4451, 1984.

[358] F. G. Stehli and S. D. Webb. *The Great American Biotic Interchange*. Plenum Press, 1985.

[359] F. G. Stehli and S. D. Webb. A kaleidoscope of plates, faunal and floral dispersals, and sea level changes. In F. G. Stehli and S. D. Webb, editors, *The Great American Biotic Interchange*, pages 3–16. Plenum Press, 1985.

[360] S. Stein, J. Engeln, D. Wiens, K. Fujita, and R. Speed. Subduction seismicity and tectonics in the Lesser Antilles arc. *Journal of Geophysical Research*, 87(B10):8642–8664, 1982.

[361] G. Sugihara. Graph theory, homology and food webs. *Proc. Symp. Appl. Math.*, 30:83–101, 1984.

[362] R. Sutton. Reinforcement learning architectures for animats. *Proceedings of the International Workshop on the Simulation of Adaptive Behavior: From Animals to Animats, MIT Press,* 1:288–296, 1991.

[363] L. K. Tanaka and S. K. Tanaka. Rainfall and seasonal changes in arthropod abundance on a tropical oceanic island. *Biotropica,* 14:114–123, 1982.

[364] M. L. Taper and T. J. Case. Models of character displacement and the theoretical robustness of taxon cycles. *Evolution,* 46:317–333, 1992.

[365] J. D. Taylor, C. J. R. Braithwaite, J. F. Peake, and E. N. Arnold. Terrestrial faunas and habitats of Aldabra during the late Pleistocene. *Phil. Trans. R. Soc. Lond. B,* 286:47–66, 1979.

[366] R. A. Thomas, P. J. Thomas, J. O Coulson, and T. Coulson. Geographic distrubution–*Anolis sagrei sagrei. Herp. Rev.,* 21:22, 1990.

[367] J. F. Tomblin. The Lesser Antilles and the Aves Ridge. In A. E. M. Nairn and F. G. Stehli, editors, *The Ocean Basins and Margins. vol. 3,* pages 467–500. The Gulf of Mexico and the Caribbean, Plenum Press, 1975.

[368] R. L. Trivers. Sexual selection and resource accruing abilities in *Anolis garmani. Evolution,* 30:253–269, 1976.

[369] G. Underwood. Revisionary notes. The anoles of the Eastern Caribbean (Sauria, Iguanidae). *Bull. Mus. Comp. Zool., Part III,* 121:191–226, 1959.

[370] R. C. Vierbuchen, Jr. The tectonics of northeastern Venezuela and the southeastern Caribbean Sea. Ph.D. thesis, Princeton University, 1979.

[371] L. von Bertalanffy. Quantitative laws in metabolism and growth. *Quarterly Review of Biology,* 32:217–231, 1957.

[372] R. B. Waide and D. Reagan. *The Food Web of a Tropical Rain Forest.* University of Chicago Press, 1993.

[373] E. E. Werner and J. F. Gilliam. The ontogenetic niche and species interactions in size-structured populations. *Annual Review of Ecology and Systematics,* 15:393–425, 1984.

[374] E. E. Werner, J. F. Gilliam, D. Hall, and G. Mittelbach. An experimental test of the effects of predation risk on habitat use in fish. *Ecology,* 64:1540–1548, 1983.

[375] S. J. Wernert, editor. *North American Wildlife.* Reader's Digest, Pleasantville, 1982.

[376] D. Westercamp, P. Andreieff, P. Bouysse, A. Mascle, and J. Baubron. Geologie de l'archipel des Grenadines (Petites Antilles meridionales). *Documents du Bureau de Recherches Geologiques et Minieres,* 92, 1985.

[377] D. Westercamp, P. Andreieff, Ph. Bouysse, A. Mascle, and J. C. Baubron. The Grenadines, southern Lesser Antilles. Part I. stratigraphy and volcano-structural evolution. In A. Mascle, editor, *Caribbean Geodynamics,* pages 109–118. Editions Technip, 27 Rue Ginoux, 75737 Paris Cedex 15, 1985.

[378] A. Wetmore. Birds of Porto Rico. *U. S. Department of Agriculture Bulletin,* 326:1–140, 1916.

[379] A. Wetmore. The birds of Porto Rico and the Virgin Islands: Psittaciformes to Passeriformes. *New York Academy of Sciences Scientific Survey of Porto Rico and the Virgin Islands,* 9:407–598, 1927.

[380] R. H. Whittaker. Evolution and measurement of species diversity. *Taxon,* 21:213–251, 1972.

[381] C. B. Williams. *Patterns in the Balance of Nature.* Academic Press, 1964.

[382] E. E. Williams. The ecology of colonization as seen in the zoogeography of anoline lizards on small islands. *Quart. Rev. Biol.,* 44:345–389, 1969.

[383] E. E. Williams. The origin of faunas. evolution of lizard congeners in a complex fauna: a trial analysis. *Evolutionary Biology,* 6:47–88, 1972.

[384] E. E. Williams. Anoles out of place: introduced anoles. In E. E. Williams, editor, *The Third Anolis Newsletter*, pages 110–118. Museum of Comparative Zoology, 1977.

[385] E. E. Williams. Ecomorphs, faunas, island size, and diverse end points in island radiations of *Anolis*. In R. B. Huey, E. R. Pianka, and T. W. Schoener, editors, *Lizard Ecology, Studies of a Model Organism*, pages 326–370. Harvard University Press, 1983.

[386] R. Williams. Simple statistical gradient-following algorithms for connectionist reinforcement learning. *Machine Learning*, 8:229–256, 1992.

[387] R. Williams and J. Peng. Reinforcement learning algorithms as function optimizers. *International Conference on Neural Networks*, 2:89–95, 1989.

[388] A. C. Wilson, S. S. Carlson, and T. J. White. Biochemical evolution. *Ann. Rev. Biochem.*, 46:573–639, 1977.

[389] A. C. Wilson and V. M. Sarich. A molecular time scale for human evolution. *Proc. Nat. Acad. Sci.*, 63:1088–1093, 1969.

[390] E. O. Wilson. The biogeography of the West Indian ants (Hymenoptera: formicidae). In J. K. Liebherr, editor, *Zoogeography of Caribbean Insects*, pages 214–230. Comstock Publishing Associates, Cornell University Press, 1988.

[391] L. D. Wilson and L. Porras. *The Ecological Impact of Man on the South Florida Herpetofauna*. University of Kansas Museum of Natural History, 1983.

[392] K. O. Winemiller. Must connectance decrease with species richness? *American Naturalist*, 134:960–968, 1989.

[393] D. Wingate. Terrestrial herpetofauna of Bermuda. *Herpetologica*, 21:202–218, 1965.

[394] J. W. Wright. The origin and evolution of the lizards of the Galapagos Islands. *TERRA*, March/April 1984:21–27, 1984.

[395] S. Y. Yang, M. Soule, and G. C. Gorman. *Anolis* lizards of the eastern Caribbean: a case study in evolution. I. Genetic relationships, phylogeny, and colonization sequence of the *roquet* group. *Syst. Zool.*, 23:387–399, 1974.

[396] P. Yodzis. The connectance of real ecosystems. *Nature*, 284:544–545, 1980.

[397] P. Yodzis. The structure of assembled communities. *J. Theoretical Biology*, 92:103–117, 1981.

[398] P. Yodzis. Energy flow and the vertical structure of real ecosystems. *Oecologia*, 65:86–88, 1984.

[399] P. Yodzis. The structure of assembled communities. II. *J. Theoretical Biology*, 107:115–126, 1984.

[400] P. Yodzis. Environment and trophodiversity. In R. E. Ricklefs and D. Schluter, editors, *Species Diversity in Ecological Communities, Historical and Geographical Perspectives*, pages 26–38. Chicago University Press, 1993.

Contents of diskette

The diskette associated with this book includes the following programs as source code:

`learn.scm` Program in Scheme that simulates rule-of-thumb learning for a forager that is minimizing time per item. One prey type is assummed with no prey escape.

`learnet.scm` Scheme program adapted from `learn.scm` to simulate rule-of-thumb learning for a forager that is maximizing energy per time. One prey type is assumed with no prey escape.

`forage.scm` Scheme program to calculate optimal foraging strategies for three prey including prey escape, and to simulate rule-of-thumb learning, assuming objective of maximizing energy per time.

`grow.scm` Scheme program to calculate optimum life history of an optimal forager.

`invade.scm` Scheme program to simulate invasion based on niche theory.

`coevolve.scm` Scheme program to simulate coevolution based on niche theory.

`foodweb.scm` Scheme program to analyze a food web and to print out summary statistics about the food web.

`foodmat.scm` Scheme program to convert a food web in matrix format to list format.

`licensus.p` Pascal program to compute Lincoln index for censusing lizards based on two distinct census days.

`licensus.c` C program translated by machine from the Pascal version.

`hrcensus.p` Pascal program to compute a lizard census from three days of censusing, using a log-linear model with an incomplete contingency table (cf. Heckel and Roughgarden [141]).

`hrcensus.c` C program translated by machine from the Pascal version.

DOS executable programs for both kinds of census are included as well.

The diskette also contains for DOS, Version 7.2 of MIT Scheme, an interpretor originally developed for UNIX workstations. The interpretor requires about 2.2Mb of space on the hard disk; the *Anolis* programs about 500Kb, for 2.7Mb total; and the interpretor runs on an 80386 or better with 4Mb or more of RAM. The interpretor will run all the Scheme programs in the book, making the diskette a complete package for verifying the models and for making modifications. The diskette is intended for readers familiar with computer programming.

Index

Note: Entries with a t. indicate tables.